Colin Ivano Cercamondi

Iron bioavailability from sorghum and millet foods

Colin Ivano Cercamondi

Iron bioavailability from sorghum and millet foods

Improving iron nutrition from sorghum and millet based diets in malaria endemic areas

Südwestdeutscher Verlag für Hochschulschriften

Impressum / Imprint
Bibliografische Information der Deutschen Nationalbibliothek: Die Deutsche Nationalbibliothek verzeichnet diese Publikation in der Deutschen Nationalbibliografie; detaillierte bibliografische Daten sind im Internet über http://dnb.d-nb.de abrufbar.
Alle in diesem Buch genannten Marken und Produktnamen unterliegen warenzeichen-, marken- oder patentrechtlichem Schutz bzw. sind Warenzeichen oder eingetragene Warenzeichen der jeweiligen Inhaber. Die Wiedergabe von Marken, Produktnamen, Gebrauchsnamen, Handelsnamen, Warenbezeichnungen u.s.w. in diesem Werk berechtigt auch ohne besondere Kennzeichnung nicht zu der Annahme, dass solche Namen im Sinne der Warenzeichen- und Markenschutzgesetzgebung als frei zu betrachten wären und daher von jedermann benutzt werden dürften.

Bibliographic information published by the Deutsche Nationalbibliothek: The Deutsche Nationalbibliothek lists this publication in the Deutsche Nationalbibliografie; detailed bibliographic data are available in the Internet at http://dnb.d-nb.de.
Any brand names and product names mentioned in this book are subject to trademark, brand or patent protection and are trademarks or registered trademarks of their respective holders. The use of brand names, product names, common names, trade names, product descriptions etc. even without a particular marking in this works is in no way to be construed to mean that such names may be regarded as unrestricted in respect of trademark and brand protection legislation and could thus be used by anyone.

Coverbild / Cover image: www.ingimage.com

Verlag / Publisher:
Südwestdeutscher Verlag für Hochschulschriften
ist ein Imprint der / is a trademark of
OmniScriptum GmbH & Co. KG
Heinrich-Böcking-Str. 6-8, 66121 Saarbrücken, Deutschland / Germany
Email: info@svh-verlag.de

Herstellung: siehe letzte Seite /
Printed at: see last page
ISBN: 978-3-8381-3564-9

Zugl. / Approved by: Zürich, ETH Zürich, Diss., 2013

Copyright © 2014 OmniScriptum GmbH & Co. KG
Alle Rechte vorbehalten. / All rights reserved. Saarbrücken 2014

Acknowledgment

I would like to thank Prof. Dr. Richard Hurrell who gave me the opportunity to carry out this PhD thesis and whose support and guidance were crucial to accomplish this thesis.

◊

I am also very grateful to Dr. Ines Egli for supervising and supporting my PhD thesis and for all her constructive contributions and the fruitful discussions throughout this thesis.

◊

I would like to thank Prof. Michael Zimmermann, M.D. who took over the guidance of my PhD thesis when Prof. Dr. Richard Hurrell became an emeritus professor.

◊

I am very grateful to Dr. Evariste Mitchikpe for his patient support in solving problems which occurred during the studies in Benin. I would like to thank Prof. Dr. Romain Dossa and Prof. Dr. Joseph Hounhouigan for their collaboration and for letting me use the facilities at the university in Cotonou, Benin.

◊

The financial contribution from the INSTAPA project which receives funding from the European Community's Seventh Framework Programme [FP7/2007–2013] under grant agreement no. 211484 is gratefully acknowledged. Furthermore this project was completed with financial support from HarvestPlus under Research Agreement 8252.

◊

Many thanks go to all my colleagues at the Human Nutrition Laboratory, ETH Zurich, Switzerland for an unforgettable time. Special thanks go to Christophe Zeder and Adam Krzystek for the time spent on isotope analyses and for their help with technical and analytical problems.

◊

Finally, huge thanks go to Sabine Yakonta and my family for being there when I needed them most!

TABLE OF CONTENTS

ABBREVIATIONS 5

SUMMARY 9

ZUSAMMENFASSUNG 15

INTRODUCTION 23

LITERATURE REVIEW 27

1   SORGHUM AND MILLETS 29
    1.1   Nomenclature of sorghum and millets 29
    1.2   Characteristics of sorghum and millet grains 31
        1.2.1   Sorghum and millet grains and their structure 31
        1.2.2   Chemical composition 33
    1.3   Growing areas and production of sorghum and millets 35
    1.4   Sorghum and millets as staple foods 38
        1.4.1   Sorghum and millet consumption 38
        1.4.2   Traditional sorghum and millet products 40
            1.4.2.1   Fermented and unfermented bread made of sorghum and millets 41
            1.4.2.2   Thick and thin porridges made of sorghum and millets 42
            1.4.2.3   Various sorghum and millet foods 44
    1.5   Prospects for a sorghum and millet food processing industry 45

2   DIETARY FACTORS AFFECTING IRON BIOAVAILABILITY FROM SORGHUM AND MILLET FOODS 48
    2.1   Iron absorption enhancers 50
    2.2   Iron absorption inhibitors 52
        2.2.1   Phytic acid 53
            2.2.1.1   Interaction of phytic acid with iron 53
            2.2.1.2   Processing steps affecting phytic acid concentration in sorghum and millet foods .. 56
        2.2.2   Polyphenols 59
            2.2.2.1   Nomenclature and occurrence in sorghum and millets 59
            2.2.2.2   Interaction of iron with polyphenols 71
            2.2.2.3   Human studies investigating the effect of polyphenols on iron absorption 73
            2.2.2.4   Improved iron bioavailability through reduced polyphenol concentration 79

3 IMPROVING IRON NUTRITION THROUGH FORTIFICATION ... 83

3.1 Conventional iron fortification ... 84

- 3.1.1 Iron compounds ... 87
  - 3.1.1.1 Water-soluble compounds ... 87
  - 3.1.1.2 Iron chelates ... 90
  - 3.1.1.3 Poorly water soluble compounds which are soluble in dilute acid ... 97
  - 3.1.1.4 Water insoluble compounds which are poorly soluble in dilute acid ... 100
  - 3.1.1.5 Encapsulated, micronized and nanostructured iron compounds ... 104
- 3.1.2 Iron-fortified complementary foods for developing countries ... 106
  - 3.1.2.1 Fortified blended foods ... 107
  - 3.1.2.2 Micronutrient powders ... 110
  - 3.1.2.3 Fortified lipid-based spreads ... 113

3.2 Biofortification ... 116

- 3.2.1 Implementation of iron biofortification ... 117
  - 3.2.1.1 Discovery phase ... 118
  - 3.2.1.2 Development phase ... 119
  - 3.2.1.3 Delivery phase ... 123
- 3.2.2 Advantages of iron biofortification ... 124
- 3.2.3 Limitations of iron biofortification ... 124
  - 3.2.3.1 General limitations of biofortification ... 124
  - 3.2.3.2 Acceptance of transgenic plants ... 125

3.3 Enhancing bioavailability of (bio)fortification iron ... 127

- 3.3.1 Adding ascorbic acid ... 127
- 3.3.2 The use of disodium EDTA ... 129
- 3.3.3 Reduction of dietary phytic acid ... 130

3.4 Iron (bio)fortification of sorghum and millets – state of the art . 132

3.5 Evaluating iron bioavailability from (bio)fortified foods ... 137

- 3.5.1 *In-vitro* methods and animal models ... 137
- 3.5.2 Human isotope studies ... 139

4 IRON NUTRITION IN AREAS WITH ENDEMIC MALARIA ... 145

4.1 Malaria ... 145

- 4.1.1 The parasite's life cycle ... 146
- 4.1.2 Asymptomatic vs. symptomatic malaria ... 148

4.2 Interaction of malaria with human iron homeostasis ... 150

- 4.2.1 Human iron homeostasis ... 151

4.2.2	Malarial anaemia	152
4.2.3	Influence of malaria on iron absorption	153
4.2.4	Malaria and non-transferrin bound iron	155
4.3	**Safety of iron interventions in malarial areas**	**158**
4.4	**Impact of malaria on the efficacy of iron fortification**	**161**

REFERENCES 163

MANUSCRIPTS 197

MANUSCRIPT 1 – MALARIA STUDY 199

MANUSCRIPT 2 – SORGHUM POLYPHENOL STUDY 225

MANUSCRIPT 3 – COMPLEMENTARY FOOD FORTIFICANT STUDY 251

MANUSCRIPT 4 – IRON-BIOFORTIFIED MILLET STUDY 275

CONCLUSIONS AND PERSPECTIVES 299

ABBREVIATIONS

A.	*Aspergillus*
AA	Ascorbic acid
AA:Fe	Ascorbic acid to iron molar ratio
ABS	Africa Biofortified Sorghum
ADI	Acceptable daily intake
BIS	Body iron stores
CFF	Complementary food fortificant
EDTA	Ethylenediaminetetraacetic acid
EFSA	European Food Safety Authority
ELISA	Enzyme-linked immunosorbent assay
DM	Dry matter
FAO	Food and Agriculture Organization
FBF	Fortified blended food
Fe^{2+}	Ferrous
Fe^{3+}	Ferric
FeBC	Ferrous bis-glycine chelate
FeFum	Ferrous fumarate
FeOP	Ferric orthophosphate
FePP	Ferric pyrophosphate
$FeSO_4$	Ferrous sulphate
FeTC	Ferric tris-glycine chelate
FM	Fresh matter
FTU	phytase unit
GAE	Gallic acid equivalent
GDF-15	Growth differentiation factor 15
G x E	Genotype and environment

ABBREVIATIONS

GM-CSF	Granulocyte macrophage colony-stimulating factor
GMO	Genetically modified organism
Hb	Haemoglobin
ICAM-1	Intercellular adhesion molecule-1
ICRISAT	International Crops Research Institute for the Semi-Arid Tropics
ID	Iron deficiency
IDA	Iron deficiency anaemia
IFN-γ	γ-interferon
JECFA	Joint Expert Committee on Food Additives of the FAO/WHO
IL	Interleukin
IP6	*Myo*- inositol- 1, 2, 3, 4, 5, 6- hexakisphophate
LNS	Lipid-based nutrient supplement
Lpa	Low phytic acid mutant
MNP	Micronutrient powder
NaFeEDTA	Sodium iron EDTA
Na_2EDTA	Disodium EDTA
NGO	Non-governmental organization
NTBI	Non-transferrin bound iron
P.	*Plasmodium*
PA	Phytic acid
PA:Fe	Phytic acid to iron molar ratio
PF	Plasma ferritin
PfEMP-1	*Plasmodium falciparum* erythrocyte membrane protein-1
PP	Polyphenol
PPO	Polyphenol oxidase
RBC	Red blood cell
RBV	Relative bioavailability

RCT	Randomized controlled trial
RES	Reticuloendothelial system
SF	Serum ferritin
TfR	Transferrin receptor
TF	Transferrin
TIBC	Total-iron-binding capacity
TNF-α	Tumor necrosis factor-α
WFP	World Food Programme
WHO	World Health Organization

SUMMARY

Background Iron deficiency (ID) without or with anaemia (IDA) is a major global health problem primarily affecting vulnerable population groups such as children <5 years of age and women of reproductive age. In the semi-arid tropics, the aetiology of IDA is multifactorial, but the major factors are low dietary iron intake and bioavailability from monotonous diets based on staple crops, such as sorghum and millets, exacerbated by chronic parasitemia such as malaria infections. Iron fortification of staple foods is considered a promising approach to prevent and correct ID in certain population groups. ID in children <5 years of age, not yet consuming significant quantities of staple foods, can be prevented by commercially iron-fortified complementary foods or by complementary food supplements also called *in-home* fortification. A more recent approach to combat ID is iron biofortification which is the development of iron-enhanced staple crops by traditional plant breeding practices and/or genetic engineering.

Sorghum and millets are important staple crops in areas of the semi-arid tropics where malaria is often endemic. Sorghum and millet foods are low in bioavailable iron and are particularly difficult to (bio)fortify with iron due to high concentrations of phytic acid (PA). In addition to PA, some sorghum and millet varieties contain considerable amounts of polyphenols (PPs) which, like PA, are known to inhibit iron absorption. The application of iron absorption enhancers, such as ascorbic acid (AA), ethylenediaminetetraacetic acid (EDTA) or enzymes degrading PA or PPs in sorghum and millet foods, potentially improves iron bioavailability from sorghum and millet foods. Malaria infections, which are known to interact with human iron homeostasis, must be considered when combatting ID in malaria endemic areas.

Aim The overall aim of this thesis was to develop approaches to improve iron nutrition from sorghum and millet based diets in malaria endemic areas. This included an evaluation of the effect of asymptomatic malaria on iron absorption, an investigation into the effect of sorghum PPs on iron absorption and the optimization of iron absorption from a newly developed complementary food fortificant (CFF) added to a thin millet gruel. Moreover, iron bioavailability and total iron absorbed from an iron-biofortified pearl millet paste was compared with that from regular-iron and post-harvest iron-fortified millet paste.

Experiments The thesis includes four investigations on iron absorption in women or young children using the stable isotope technique. Three investigations were conducted in Benin with women or children and one in Switzerland with women.

Manuscript 1: Iron absorption and utilization from an iron-fortified sorghum porridge (3 mg iron) were estimated by using oral and intravenous isotope labels in 23 afebrile Beninese women with a positive malaria smear (asexual *P. falciparum* parasitemia; >500 parasites/µL blood). The women were studied while infected, treated, and then restudied 10 days after treatment. Iron status, hepcidin, and inflammation indexes were measured before and after treatment.

Manuscript 2: Three iron absorption studies were conducted to investigate the inhibiting effects of sorghum PPs on iron absorption and the potential enhancing effects of AA, sodium iron EDTA (NaFeEDTA) and polyphenol oxidase treatment using laccase. Studies were conducted in 50 women residing in Switzerland. They consumed dephytinized test meals based on white and brown sorghum varieties with different PP concentrations.

Manuscript 3: Optimization of iron bioavailability from a CFF, primarily designed to prevent iron deficiency, was investigated in three absorption studies. Fifty-two young Beninese children consumed test meals consisting of 80 g millet porridge mixed with 30 g CFF and 40 mg AA. Study 1 compared iron absorption from meals fortified with 6 mg iron as ferrous sulphate ($FeSO_4$) with meals fortified with 3 mg iron as $FeSO_4$ and 3 mg iron as NaFeEDTA. Study 2 compared iron absorption from $FeSO_4$-fortifed meals (6 mg iron) without and with 40 mg extra AA (total AA = 80 mg). Study 3 compared iron absorption from $FeSO_4$-fortified meals (6 mg iron) with meals containing phytase added prior to consumption once without and once with extra AA.

Manuscript 4: A stable isotope study in 20 Beninese women (plasma ferritin <25 µg/L) using a multiple meal design was carried out to compare iron bioavailability and total iron absorbed from regular-iron, iron-biofortified and post-harvest iron-fortified pearl millet. The three different composite test meals consisted of millet paste either based on regular-iron (1.5 mg iron/serving), iron-biofortified (5.5 mg iron/serving) or post-harvest iron-fortified (5.2 mg iron/serving) pearl millet accompanied by a leafy vegetable sauce or by an okra sauce. Each test meal was consumed 10 times over 5 days (2 portions/day). Test meal servings were labelled with an extrinsic iron isotope tag of 0.4 mg.

Results   *Manuscript 1:* Clearance of asymptomatic malaria parasitemia increased dietary iron absorption (from 10.2% to 17.6%; $P < 0.01$) but did not affect systemic iron utilization (85.0% compared with 83.1%; NS). Malaria treatment reduced low-grade inflammation, as reflected by decreases in serum ferritin, C-reactive protein and interleukin-6, -8, -10 ($P < 0.05$); this was accompanied by a reduction in serum hepcidin of approximately 50%, from 2.7 to 1.4 nmol/L ($P < 0.01$). Treatment decreased serum erythropoietin and growth differentiation factor 15 ($P < 0.05$).

Manuscript 2: In study 1, iron absorption from sorghum meals with 17 mg PPs (8.5%) was higher than from sorghum meals with 73 mg PPs (3.2%) and 167 mg PPs (2.7%) ($P < 0.001$). Meals containing 73 and 167 mg PPs showed no difference in absorption ($P = 0.9$). In study 2, iron absorption from meals (167 mg PPs) fortified with NaFeEDTA (4.6%) was higher than from the same meals fortified with $FeSO_4$ (2.7%; $P < 0.001$) but lower compared with $FeSO_4$-fortified meals containing 17 mg PPs (10.7%; $P < 0.001$). In study 3, PP reduction by laccase (from 167 to 42 mg PPs/meal) did not improve iron absorption compared with meals with 167 mg PPs (4.8% vs. 4.6%; $P = 0.4$). Adding AA increased iron absorption to 13.6% ($P < 0.001$).

Manuscript 3: In study 1, iron absorption was higher from $FeSO_4$ (8.4%) than from the mixture of NaFeEDTA and $FeSO_4$ (5.9%; $P < 0.01$). In study 2, adding extra AA to the CFF mixed with millet porridge increased absorption (11.6%) compared with standard AA concentration (7.3%; $P < 0.001$). In study 3, absorption from test meals containing phytase without or with extra AA (15.8 and 19.9%, respectively) was increased compared with test meals without phytase (8.0%; $P < 0.001$). The addition of extra AA to meals containing phytase increased absorption when compared with the test meals containing phytase without extra AA ($P < 0.05$).

Manuscript 4: Fractional iron absorption from the test meals based on regular-iron millet did not differ compared with iron-biofortified millet meals (7.5% vs. 7.5%; $P = 1.0$) resulting in a higher quantity of total iron absorbed from the iron-biofortified millet meals (527 µg vs. 1125 µg; $P < 0.001$). Fractional iron absorption from post-harvest iron-fortified millet (10.4%) meals was higher than from regular-iron and iron-biofortified millet meals ($P < 0.01$). Total iron absorbed from the test meals based on post-harvest iron-fortified millet (1500 µg) was higher than from the regular-iron and iron-biofortified millet meals ($P < 0.001$ and $P < 0.01$, respectively).

Conclusions The findings in *manuscript 1* show that asymptomatic malaria parasitemia decreases dietary iron absorption but does not influence systemic iron utilization. The negative effect is mediated through low-grade inflammation and it may contribute to ID and IDA or may blunt efficacy of fortification programmes in malaria-endemic areas.

Results of *manuscript 2* demonstrate the strong inhibition of PPs from brown sorghum on iron absorption. The results suggest that dephytinization of sorghum-based foods alone would not improve iron absorption if foods contain considerable amounts of PPs from brown sorghum. Furthermore, they show that especially AA and to a lesser extent NaFeEDTA but not laccase are promising enhancers to improve iron bioavailability from foods containing inhibitory PPs from brown sorghum.

As shown in *manuscript 3*, iron absorption from a lipid-based complementary food fortificant mixed with cereal porridge can be improved by using phytase and AA but not by using a mixture of $FeSO_4$ and NaFeEDTA. Based on the results, the use of a $FeSO_4$/NaFeEDTA mixture as iron fortificant in a lipid-based complementary food cannot be recommended. The addition of phytase and high concentrations of AA is the most promising approach to provide adequate bioavailable iron to young children via a CFF which also delivers energy and protein.

In *manuscript 4*, the findings show that the total amount of iron absorbed from meals based on iron-biofortified pearl millet is about twice of that from regular-iron millet meals and approximately 2/3 from that of post-harvest iron-fortified millet meals containing the same amount of iron. This indicates that iron biofortification of millet is a promising approach to provide additional bioavailable iron in the diets of millet consuming communities with limited access to conventionally fortified foods and may help to combat ID.

The findings in this thesis demonstrate that improvement of iron nutrition in sorghum and millet consuming communities is challenging because of the negative influence of PA, PPs and malaria. Nevertheless, it can be improved by using iron absorption enhancers in future fortification programs of sorghum flour (e.g. AA or NaFeEDTA) or in lipid-based iron-fortified supplements (e.g. phytase and AA), and also by iron-biofortified millets providing additional bioavailable iron into the diet.

ZUSAMMENFASSUNG

Hintergrund Eisenmangel mit oder ohne Anämie ist ein weltweit weitverbreitetes Gesundheitsproblem, das vor allem Kinder unter fünf Jahren und Frauen im gebärfähigen Alter betrifft. In semi-ariden tropischen Gebieten hat Eisenmangelanämie verschiedene Ursachen, jedoch gelten die geringe Aufnahme von bioverfügbarem Eisen bei monotonen Ernährungsgewohnheiten basierend auf Grundnahrungsmitteln wie z. B. Sorghumhirse oder anderen Hirsearten, und der gleichzeitige Einfluss von chronischen Infektionskrankheiten, wie z. B. Malaria, als die primären Ursachen. Ein vielversprechender Ansatz für die Korrektur von Eisenmangel ist die Eisenanreicherung von Grundnahrungsmitteln. Bei Kleinkindern, die erst geringe Mengen an Grundnahrungsmitteln verzehren, kann Eisenmangel durch den Konsum von kommerzieller mit Eisen angereicherter Beikost oder durch die Zugabe von Supplementen zur selbstgemachten Beikost korrigiert werden. Ein neuerer Ansatz ist die Biofortifizierung von Grundnahrungspflanzen, bei welcher die biosynthetischen Kapazitäten einer Pflanze genutzt werden, um ihren Eisengehalt zu steigern. Biofortifizierung kann mittels Gentechnik und/oder klassischer Züchtung erfolgen.

Sorghumhirse und andere Hirsearten sind wichtige Grundnahrungsmittel in den semi-ariden tropischen Gebieten, in welchen Malaria oft endemisch auftritt. Lebensmittel basierend auf Hirse beinhalten nur geringe Mengen an bioverfügbarem Eisen und die konventionelle Eisenanreicherung nach der Ernte oder die Biofortifizierung sind aufgrund des relativ hohen Phytinsäuregehalts schwierig. Einige Hirsearten enthalten zudem Polyphenole, welche wie Phytinsäure die Eisenaufnahme beim Menschen inhibieren. Der Einsatz von Eisen-Absorptionsverstärkern wie z. B.

Ascorbinsäure, Ethylendiamin-tetraessigsäure (EDTA) oder Enzymen, welche Phytinsäure oder Polyphenole abbauen, kann möglicherweise die Eisenabsorption aus Lebensmitteln basierend auf Hirse erhöhen. Bei der Korrektur von Eisenmangel in endemischen Malariagebieten muss zusätzlich auch der Einfluss von Malaria auf die menschliche Eisen-Homöostase berücksichtigt werden.

Zielsetzung Das Hauptziel der Arbeit war die Entwicklung von Ansätzen zur Verbesserung der Eisenaufnahme aus Lebensmitteln basierend auf Sorghumhirse oder anderen Hirsearten in Gebieten mit endemischer Malaria. Der Fokus lag dabei auf der Evaluierung des Effektes von asymptomatischer Malaria auf die Eisenabsorption, der Erforschung des Effektes von Sorghumhirse-Polyphenolen auf die Eisenabsorption und der Optimierung der Eisenabsorption aus einem lipidhaltigen Beikost-Supplement. Darüber hinaus wurde gewöhnliches, biofortifiziertes und konventionell angereichertes Perlhirsen-Mehl auf die Bioverfügbarkeit von Eisen sowie auf die absorbierte Gesamtmenge von Eisen getestet.

Experimente Die Arbeit enthält vier Untersuchungen, welche die Eisenabsorption bei Frauen oder Kleinkindern mittels stabilen Eisenisotopen bestimmen. Drei Untersuchungen wurden in Benin durchgeführt, eine in der Schweiz.

Manuskript 1: Mittels oraler und intravenöser Verabreichung von Eisenisotopen wurde die Eisenabsorption und Eiseninkorporation aus einem mit 3 mg Eisen angereicherten Sorghum-Brei untersucht. Die Studie wurde mit 23 beninischen Frauen mit asymptomatischer (fieberloser) Malaria durchgeführt (asexuelle *P. falciparum* Parasitämie; >500 Parasiten/µL Blut). Die Untersuchungen wurden während der Infektion und dann erneut 10 Tage nach Infektionsbehandlung durchgeführt. Eisenstatus, Hepcidin und Entzündungsparameter wurden bei den Frauen vor und nach der Behandlung gemessen.

Manuskript 2: In drei Eisenabsorptionsstudien wurde der inhibierende Effekt von Sorghum-Polyphenolen auf die Eisenabsorption erforscht sowie Möglichkeiten untersucht, diesem Effekt mit der Anwendung von Ascorbinsäure, Natriumeisen-EDTA (NaFeEDTA) oder Laccase entgegenzuwirken. In den drei Studien konsumierten 50 in der Schweiz wohnhafte Frauen dephytinisierte Sorghumhirse-Testmahlzeiten mit verschiedenen Polyphenol-Konzentrationen.

Manuskripts 3: Die Optimierung der Bioverfügbarkeit von Eisen aus einem eigens für die Prävention von Eisenmangel angereicherten Beikost-Supplement wurde mit stabilen Isotopen in drei Studien untersucht. Diese Studien wurden mit 52 beninischen Kleinkindern durchgeführt und die Testmahlzeit bestand aus 80 g Hirsebrei gemischt mit dem Beikost-Supplement sowie 40 mg Ascorbinsäure. In Studie 1 wurde die Eisenabsorption aus einer fortifizierten Testmahlzeit mit 6 mg Eisen in Form von Eisensulfat ($FeSO_4$) mit der aus einer mit 3 mg Eisen in Form von $FeSO_4$ und 3 mg Eisen in Form von NaFeEDTA fortifizierten Testmahlzeit verglichen. In der 2. Studie wurde die Eisenabsorption aus $FeSO_4$-fortifizierten Testmahlzeiten (6 mg Eisen) einmal ohne (40 mg) und einmal mit zusätzlicher Ascorbinsäure (80 mg) gegenübergestellt. In Studie 3 wurde eine $FeSO_4$-fortifizierte Testmahlzeit (6 mg Eisen) mit phytasehaltigen Testmahlzeiten verglichen, welche einmal ohne (40 mg) und einmal mit zusätzlicher Ascorbinsäure (80 mg) versetzt waren.

Manuskript 4: In einer Studie mit stabilen Isotopen wurde die relative Eisenabsorption sowie die Gesamtmenge an aufgenommenem Eisen aus Mahlzeiten verglichen, welche auf gewöhnlicher Perlhirse, einer mit Eisen biofortifizierten Perlhirse oder einer konventionell mit Eisen angereicherten Perlhirse basierten. Zwanzig beninische Frauen (Plasma-Ferritin <25 µg/L) konsumierten während fünfzehn Tagen drei verschiedene Testmahlzeiten (jeweils 2 Portionen/Tag während 5 Tagen/Testmahlzeit) bestehend aus einer Hirse-Polenta zusammen mit einer Blattgemüsesauce oder einer Okra-Sauce. Die drei verschiedenen Testmahlzeiten basierten jeweils auf der gewöhnlichen Hirse, der biofortifizierten Hirse oder der konventionell angereicherten Hirse. Alle Testmahlzeit-Portionen wurden mit 0.4 mg Eisenisotopen extrinsisch markiert.

Resultate *Manuskript 1:* Nach Behandlung der asymptomatischen Malaria stieg die Eisenabsorption von 10.2% auf 17.6% ($P < 0.01$). Die Eiseninkorporation blieb nach der Behandlung gleich (85.0% vs. 83.1%; NS). Die Behandlung führte dazu, dass die leicht erhöhten Entzündungsparameter Serum-Ferritin, C-reaktives Protein sowie Interleukin-6, -8, -10 abnahmen ($P < 0.05$). Darüber hinaus gingen nach der Behandlung die Werte für Serum-Erythropoetin, "Growth Differentation" Faktor 15 ($P < 0.05$) sowie für Hepcidin ($P < 0.01$) zurück. Letzteres nahm nach der Behandlung um zirka 50% ab, von 2.7 auf 1.4 nmol/L.

Manuskript 2: In der 1. Studie war die Eisenabsorption aus den Sorghumhirse-Mahlzeiten mit 17 mg Polyphenolen (8.5%) höher als aus jenen Mahlzeiten mit 73 mg (3.2%) und 167 mg (2.7%) Polyphenolen ($P < 0.001$). Bei den Eisenabsorptionen aus den Mahlzeiten mit 73 mg und 167 mg Polyphenolen konnte kein Unterschied festgestellt werden ($P = 0.9$). In Studie 2 war die Eisenabsorption aus einer mit NaFeEDTA-fortifizierten Sorghumhirse-Mahlzeit mit 167 mg Polyphenolen (4.6%) höher als aus derselben Mahlzeit fortifiziert mit $FeSO_4$ (2.7%; $P < 0.001$). Die Absorption aus der NaFeEDTA-fortifizierten Mahlzeit war aber immer noch geringer als jene aus der FeSO$_4$-fortifizierten Mahlzeit mit nur 17 mg Polyphenolen (10.7%; $P < 0.001$). In Studie 3 führte der Polyphenol-Abbau mittels Laccase auf 42 mg/Testmahlzeit zu keiner verbesserten Eisenabsorption im Vergleich zu der Mahlzeit mit 167 mg Polyphenolen (4.8% vs. 4.6%; $P = 0.4$). Die Zugabe von Ascorbinsäure erhöhte die Eisenabsorption (13.6%; $P < 0.001$).

Manuskript 3: In Studie 1 wurde FeSO$_4$ (8.4%) besser absorbiert als die Mischung aus FeSO$_4$ und NaFeEDTA (5.9%; $P < 0.01$). Das Beimischen von zusätzlicher Ascorbinsäure zu den Testmahlzeiten in Studie 2 führte zu einer höheren Eisenabsorption im Vergleich zu Testmahlzeiten, welche die Standard-Konzentration an Ascorbinsäure enthielten (11.6% vs. 7.6%; $P < 0.001$). In der 3. Studie war die Eisenabsorption aus Testmahlzeiten gemischt mit Phytase sowohl ohne (15.8%) als auch mit zusätzlicher Ascorbinsäure (19.9%) höher als die Eisenabsorption aus Testmahlzeiten ohne Phytase (8.0%; $P < 0.001$). Die Eisenabsorption aus phytasehaltigen Testmahlzeiten mit zusätzlicher Ascorbinsäure war höher als jene aus phytasehaltigen Testmahlzeiten ohne zusätzliche Ascorbinsäure ($P < 0.05$).

Manuskript 4: Die fraktionelle Eisenabsorption aus den Mahlzeiten basierend auf der Hirse mit gewöhnlichem Eisengehalt (7.5%) unterschied sich nicht von der fraktionellen Eisenabsorption aus den Mahlzeiten, welche mit der biofortifizierten Hirse zubereitet wurden (7.5%; $P = 1.0$). Die absolute Menge an aufgenommen Eisen pro Studientag aus den Mahlzeiten mit biofortifizierter Hirse (1125 µg) war höher als aus den Mahlzeiten zubereitet mit der gewöhnlichen Hirse (527 µg; $P < 0.001$). Die fraktionelle Eisenabsorption aus den Mahlzeiten, welche auf der konventionell angereicherter Hirse basierten (10.4%), war höher verglichen mit den Mahlzeiten zubereitet mit der biofortifizierten Hirse und den

Mahlzeiten basierend auf der gewöhnlichen Hirse ($P < 0.01$). Die totale Menge an aufgenommenem Eisen pro Studientag aus den Mahlzeiten mit der konventionell angereicherten Hirse (1500 µg) war höher als aus den Mahlzeiten zubereitet aus der biofortifizierten Hirse ($P < 0.01$) und aus der gewöhnlichen Hirse ($P < 0.001$).

Schlussfolgerungen Die im *Manuskript 1* beschriebene Studie lässt erkennen, dass asymptomatische Malariainfektionen die Eisenabsorption im menschlichen Darm inhibieren, aber keinen Einfluss auf die systemische Verwertung von Eisen haben. Diese Inhibition, welche durch Inflammationsprozesse im menschlichen Körper gesteuert wird, führt möglicherweise zu einem vermehrten Vorkommen von Eisenmangel und Eisenmangelanämie in Regionen mit endemischer Malaria und einer verminderten Wirksamkeit von mit Eisen angereicherten Lebensmitteln.

Die Untersuchungen im *Manuskript 2* zeigen, dass Polyphenole von brauner Sorghumhirse die Eisenabsorption stark inhibieren. Die Resultate weisen darauf hin, dass die Dephytinisierung von Lebensmitteln aus brauner Sorghumhirse nicht ausreichen würde, um die Eisenabsorption daraus zu erhöhen. Die Untersuchungen zeigen auch, dass vor allem Ascorbinsäure und in geringerem Ausmass NaFeEDTA dem inhibierenden Einfluss der Polyphenole entgegenwirken. Im Gegensatz dazu hatte der partielle Abbau der Polyphenole durch Laccase keine Verbesserung der Eisenabsorption zur Folge.

Die Studien im *Manuskript 3* legen dar, dass der Einsatz von Phytase oder Ascorbinsäure die Bioverfügbarkeit von Eisen aus einem Gemisch aus Getreidebrei und lipid-haltigem Beikost-Supplement optimieren kann. Die Zugabe von Eisen als NaFeEDTA im Beikost-Supplement führte nicht zu einer verbesserten Bioverfügbarkeit und kann deshalb momentan nicht empfohlen werden. Vor allem der gleichzeitige Gebrauch von Phytase und Ascorbinsäure in einem mit Eisen angereicherten lipid-haltigem Beikost-Supplement ist ein vielversprechender Ansatz um die Menge an bioverfügbarem Eisen in der Beikost zu erhöhen und gleichzeitig auch die Energie- und Proteinzufuhr beim Kleinkind zu steigern.

Die Resultate im *Manuskript 4* zeigen, dass die absolute Menge an aufgenommenem Eisen aus den mit biofortifizierter Hirse zubereiteten Mahlzeiten doppelt so hoch ist, wie jene aus den Mahlzeiten zubereitet aus der gewöhnlichen Hirse. Die absolute Menge an aufgenommenem Eisen aus den mit biofortifizierter Hirse zubereiteten Mahlzeiten betrug ungefähr 2/3 von jener aus Mahlzeiten basierend auf konventionell angereicherter Hirse. Diese Resultate lassen erkennen, dass eine mit Eisen biofortifizierte Hirse jene Populationsgruppen, welche Hirse als Grundnahrungsmittel verwenden, mit mehr bioverfügbarem Eisen versorgt als die gewöhnliche Hirse. Der Einsatz von biofortifizierter Hirse ist deshalb ein vielversprechender Ansatz zur Korrektur und Prävention von Eisenmangel bei Populationsgruppen mit limitiertem Zugang zu konventionell mit Eisen angereicherten Lebensmitteln.

Die Zusammenfassung aller Studienresultate zeigt, dass eine Verbesserung der Eisenaufnahme aus Lebensmitteln basierend auf Sorghumhirse oder anderen Hirsearten nur unter Berücksichtigung der negativen Einflüsse von Phytinsäure, Polyphenolen und endemischer Malaria möglich ist. Die Arbeit zeigt weiter, dass der Gebrauch von Eisenabsorption-Verstärkern in Hirsemehlen oder Beikost-Supplementen angereichert mit Eisen oder der Konsum von biofortifizierten Hirsearten wertvolle Ansätze sind um die Einnahme von bioverfügbarem Eisen zu erhöhen.

INTRODUCTION

Iron deficiency with or without anaemia is common in sub-Saharan Africa, particularly in children <5 years of age and in women of childbearing age (World Health Organization 2001). There is compelling evidence that infants and young children with iron deficiency anaemia are at risk for impaired cognitive, motor, social-emotional, and neurophysiological development in the short- and long-term outcome (Lozoff 2007). In women of childbearing age, iron deficiency anaemia is associated with increased maternal and perinatal mortality, and reduced work capacity (Zimmermann and Hurrell 2007). The aetiology of iron deficiency anaemia in developing countries is multifactorial, but the major causes are low dietary iron intake and bioavailability from monotonous cereal-based diets (Zimmermann et al. 2005a), aggravated by chronic parasitic infections such as malaria (Newton et al. 1997).

Sorghum and millets are main staple foods for many people living in developing countries of the semi-arid tropics. Their importance is based on their ability to adequately yield under harsh climate conditions. In case of future climate change, drought resistant sorghum and millets may become even more important. Sorghum and millets are frequently consumed as thin or thick porridges and as fermented breads. The thin porridges based on sorghum and millets are popular complementary foods in many sub-Saharan African countries and in India (FAO 1995). Post-harvest fortification of cereal flours with iron is the main strategy to combat iron deficiency in communities with cereal staple foods. A more novel approach is iron biofortification which is the development of crops with increased concentrations of native iron. Fortification or biofortification of cereals, such as sorghum or millets, however, is particularly difficult, since they contain significant amounts of phytic acid and in addition can contain polyphenols. Both are well-known inhibitors of human iron absorption (Hurrell et al. 2003). Iron fortification and biofortification of sorghum and millets is in the early stages and research on the iron bioavailability from post-harvest iron-fortified and iron-biofortified sorghum and millets is required to move things forward.

In-home fortification with complementary food supplements, such as lipid-based spreads or micronutrient powders, is a promising approach to reach infants and young children in rural areas in developing countries where availability of commercially produced iron-fortified complementary foods is limited. Lipid-based spreads have the advantage that they provide macronutrients, such as fat (for energy and essential fatty acids) and protein, in addition to micronutrients. In response to the mounting evidence that untargeted high iron doses consumed in areas with high malaria transmission lead to increased child morbidity and mortality, iron concentrations in complementary food supplements to be consumed in such areas can be adjusted to levels which are judged safe. However, research is needed on the interaction of malaria with human iron homeostasis and its potential negative impact on the efficacy of iron fortification programs (Hurrell 2010a).

The present thesis was part of the INSTAPA project, which aims to identify novel staple food-based approaches to improve micronutrient malnutrition for better health and development of women and children in sub-Saharan Africa. The thesis consists of a literature review and four manuscripts. The first part of the review deals with sorghum and millets. The second part is focused on dietary factors that affect iron bioavailability from foods made of sorghum and millets. Then, fortification as a strategy to improve iron nutrition is reviewed. Lastly, the role of iron nutrition in malaria endemic areas is discussed including a brief overview on human iron homeostasis. The first manuscript describes the effect of asymptomatic malaria on iron absorption and utilization by Beninese women from iron-fortified sorghum porridge. The second manuscript reports a series of iron absorption studies that investigated the inhibitory effect of sorghum polyphenols on human iron absorption and potential approaches to counteract these effects. The third manuscript deals with the optimization of iron bioavailability from a lipid-based complementary food fortificant designed to prevent iron deficiency in young children. Finally, the fourth manuscript compares iron absorption in young women from regular-iron, iron-biofortified and post-harvest iron-fortified pearl millet. All iron absorption studies were done using stable isotope technique in groups of women or children either in Benin or Switzerland.

LITERATURE REVIEW

1 SORGHUM AND MILLETS

After wheat, rice, maize and barley, sorghum is the fifth most important cereal crop in the world in terms of production and growing area. Millets, a collective term for a number of small seeded grass grain crops, are the world's seventh most important cereal grains (FAO 1995). Although these cereals have been important staples for many centuries, especially for millions of underprivileged people in the semi-arid tropics, they have not received the same attention in science and development as other cereals (Vietmeyer and Ruskin 1996). Nowadays, sorghum and millets are still grown by mainly small-holder cultivators and they are usually not traded in inter- or national markets. Not surprisingly, the foregoing negligence of sorghum and millets resulted in numerous common and vernacular names by which they are known (ICRISAT/FAO 1996). The following chapter briefly explains this complex nomenclature and describes the characteristics of sorghum and millet grains, their growing areas including production and consumption, their processing into traditional foods and their potential for the food processing industry.

1.1 Nomenclature of sorghum and millets

The origin of the name "sorghum" remains unclear. Most likely it derives from the Latin verb "surgere" meaning "to rise". Regarding millets, it is reasonably assured that the term "millet" originates from the Latin word "millesimum" meaning "a thousandth part". The diminutive "mil" is often used to describe something which is extremely small which fits perfectly with the term "millet" including many different cereal species characterized by their small grains (Hulse et al. 1980). All these species have various common names according to the lingua franca or the place of cultivation. The biological and common names of sorghum and millets are summarized in Table 1. The genus sorghum comprises numerous species, but *Sorghum (S.) bicolor* is the primarily cultivated specie (FAO 1995). It has to be stressed that most of the ancient French language literature used the name *Sorghum vulgare* instead of *S. bicolor* (Hulse et al. 1980). *S. bicolor* is divided into 5 races: 1) *bicolor* found in Africa and Asia, 2) *guinea* typically grown in Western Africa, 3) *caudatum* particularly found in Central Africa, 4) *kafir* mainly planted in Eastern Africa from Tanzania southward, and 5) *durra* found in Northern and Eastern Africa, Orient and India (Harlan and Dewet 1972).

Millets are classified in two broad categories: pearl millet (*Pennisetum glaucum*) and the small or minor millets (Belton and Taylor 2002). Pearl millet represents a separate category because it is the most extensively cultivated millet specie worldwide and especially widespread in Western African countries and India (Vietmeyer and Ruskin 1996). Minor millets include finger millet (*Eleusine coracana*), foxtail millet (*Setaria italica*), proso millet (*Panicum miliaceum*), little millet (*Panicum sumatrense*), barnyard millet (*Echinochloa crus-galli*), kodo millet (*Paspalum scrobiculatum*), fonio (*Digitaris exilis* and *Digitaris iburua*), Job's tears (*Coix lachrymal-jobi*) and Teff (*Eragrostis tef*) (Hulse et al. 1980; Belton and Taylor 2002). Fonio and job's tears are of marginal local importance and teff is botanically not strictly a millet, but because of the very small grains and the extremely important status in Ethiopia, it is usually referred to as millet (Hulse et al. 1980; FAO 1995). Within the minor millets, finger millet is the most important specie widely grown in Eastern and Southern Africa, and as well India (FAO 1995).

Table 1: Biological and common names of sorghum and millets (modified from FAO 1995).

Biological name	Common names
Sorghum bicolor	Sorghum, great millet, guinea corn, kafir corn, aura, mtama, jowar, cholam. kaoliang, milo, milo-maize
Pennisetum glaucum	Pearl millet, cumbu, spiked millet, bajra, bulrush millet, candle millet, dark millet
Eleusine coracana	Finger millet, African millet, koracan, ragi, wimbi, bulo, telebun
Setaria italica	Foxtail millet, Italian millet, German millet, Hungarian millet, Siberian millet
Panicum miliaceum	Proso millet, common millet, hog millet, broom-corn millet, Russian millet, brown corn
Panicum sumatrense	Little millet
Echinochloa crus-galli	Barnyard millet, sawa millet, Japanese barnyard millet
Paspalum scrobiculatum	Kodo millet
Digitaris exilis/iburua[1]	Fonio, hungry rice, findi, fundi
Coix lachrymal-jobi	Job's tears
Eragrostis tef[2]	Teff

[1] Kuta et al. (2003).
[2] Botanically not strictly a millet but usually referred to as millet.

1.2 Characteristics of sorghum and millet grains

1.2.1 Sorghum and millet grains and their structure

Like all the staple food grains grown around the world, sorghum and millets belong to the grass family *Poaceae*. Depending on the specie and the genetic diversity, the appearance of the sorghum and millet plants, especially the panicles and grains, can be remarkably different in colour, shape, size, and other certain anatomical components (FAO 1995). Nevertheless, the basic grain structure of sorghum and different millets is similar. The principle structural components are bran (pericarp + aleurone layer), germ (embryo + scutellum) and endosperm (Figure 1) (Hama et al.

2011). Endosperm is the largest fraction, which represents about 82% of the grain weight in sorghum and about 75% in millets. The germ stands for about 10% and 17% of the grain weight in sorghum and millets, respectively; bran represents about 8% in both sorghum and millets (Abdelrahman et al. 1984; Favier 1989).

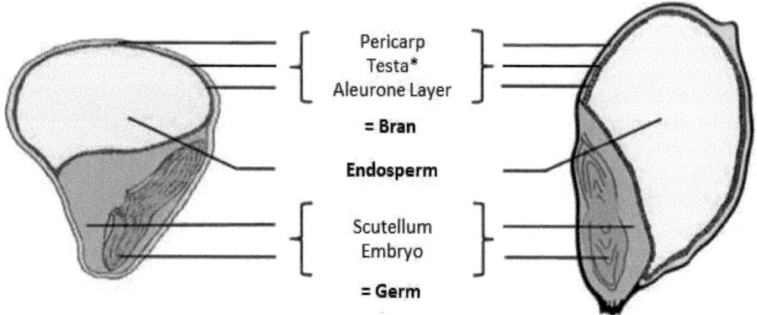

Figure 1: Basic anatomical components of millet (left side) and sorghum (right side) grains (modified from Hama et al. 2011). *Testa is the seed-coat underneath the pericarp. The thickness of the testa varies and some genotypes have a partial testa, while in others it is not apparent or absent (FAO 1995).

Grains of sorghum and pearl millet belong to the caryopsis type, where the pericarp is entirely fused to the endosperm. Other millets, such as finger millet, are of the utricle type where the pericarp is just loosely attached to the endosperm at only one point. Therefore, the pericarp can easily break away and leave the testa (seed-coat underneath the pericarp) to protect the inner endosperm. The weight of 1000 sorghum grains is about 25–30 g. They are spherical and due to a pigmented pericarp containing coloured phenolic compounds, they can appear white, yellowish, red, or brown. Pearl millet grains can appear grey, white, lemon yellow, brown and purple. In addition to pigments located in the pericarp, also the testa is sometimes pigmented in sorghum and millets. The shape of pearl millets is irregular; it can be ovoid, hexagonal or globose. Pearl millet grains are lighter than sorghum grains. The average weight of 1000 pearl millet grains is 8 g. Compared with other millets the size of pearl millet grains is much bigger (FAO 1995). The grain size of teff and fonio is extremely small, with 1000 grains weighing about 0.3–0.4 g and 0.5 g, respectively (Almgard 1963; Belton and Taylor 2002). Table 2 summarizes a few structural traits of grains of sorghum and a few millets.

Table 2: Structural traits of grains of sorghum and a few millets (modified from FAO 1995).

Grain	Type	Shape	Colour	1 000-grain weight (g)
Sorghum	Caryopsis	Spherical	White, yellowish, red, brown	25–30
Pearl millet	Caryopsis	Ovoid, hexagonal, Globose	Grey, white, brown lemon yellow, purple	2.5–14
Finger millet	Utricle	Globose	Yellowish, white, red, brown, violet	2.6
Teff[1]	Caryopsis	Globose	White, brown	0.3–0.4

[1] Almgard (1963).

1.2.2 Chemical composition

This section summarizes carbohydrates, proteins, lipids and vitamins in sorghum and millet. Iron, phytic acid (PA) and polyphenols (PPs) are discussed in separate sections (see chapter 2: Dietary Factors Affecting Iron Bioavailability from Sorghum and Millet Foods).

Carbohydrate concentration in sorghum and millets is similar, but tends to be lower in some millets such as little or barnyard millet. The main storage form of carbohydrate in sorghum and millets is starch. Soluble sugar, pentosans, cellulose, and hemicellulose are usually low. In sorghum types with regular endosperm, amylose concentration is between 20 to 30%, while the starch of some waxy varieties is practically 100% amylopectin (FAO 1995). Protein is the second main component in sorghum and millets. The protein concentration (10–13%) and composition is largely affected by genotype, water availability, temperature, soil fertility, and environmental conditions during grain development (Hulse et al. 1980). While the major protein fractions in millets are prolamins, glutelins are the major fractions in sorghum (FAO 1995). Due to the characteristic lack of gluten, sorghum and millet are not suitable for raised bread (Chiba et al. 2012), on the other hand, they represent a food source for the millions of people who are intolerant to gluten (celiac disease) (Cureton and Fasana 2009). Regarding their essential amino acid profile, millets generally have the better chemical score than sorghum. However, the shared feature is that lysine, methionine, cysteine and tryptophan are always found to be the most limiting amino acids (FAO 1995). The total fat concentration varies among millets and is highest in little and pearl millet with about 5%. In sorghum, the total fat concentration is reported to be around 3% (Hulse et al. 1980). The lipid fraction is almost entirely located in the germ and aleurone layer. Eighty percentage of the total fat is found in the germ (FAO 1995). With high concentrations of linoleic (49%), oleic (31%) and palmitic acids (14%), the fatty acid composition of sorghum is similar to that of maize but slightly more unsaturated (Rooney 1978).

Sorghum and millets are generally good sources of B and E vitamins except B12 (FAO 1995). However, niacin bioavailability in cereal grains is possibly limited (Wall and Carpenter 1988). Mature sorghum and millet grains do not contain Vitamin C (FAO 1995). Sorghum varieties with yellow-endosperm contain ß-carotene which can be converted to vitamin A by the human body (Blessin et al. 1958), but are of little importance as a dietary source of vitamin A precursor (FAO 1995). All the vitamins are concentrated in the germ and aleurone layers of the grain. Processing steps, such as decortication, lead to a removal of these tissues (Nambiar et al. 2011).

1.3 Growing areas and production of sorghum and millets

The worldwide relative importance of sorghum and millets in terms of percentage sorghum and millets growing area of national arable area is shown in Figure 2 and Figure 3. Approximately 90% of the world's sorghum area and 95% of the world's millet area are found in the developing countries, mainly in Africa and Asia (ICRISAT/FAO 1996). In industrialized countries, sorghum and millets are almost exclusively cultivated for animal feeding. The growing areas of sorghum and millets cultivated as food crops are in most instances in the arid and semi-arid tropics. The precise definition of "semi-arid" is difficult, but in general, it describes those regions where evapo-transpiration exceeds rainfall for more than half of the year. Almost all African countries surrounding the Sahara, much of Middle and Eastern Africa, a huge area of central India and some parts of Southeast Asia and South America can be classified as semi-arid tropics (Hulse et al. 1980). The ability to yield under harsh conditions with limited water resources and without application of any fertilizers makes sorghum and millets particularly suitable for such areas (Vietmeyer and Ruskin 1996). In respect of increasing temperature due to climate change having severe effects on the water balance in semi-arid and arid areas, importance of sorghum and millets may increase in such areas. Sorghum and millets were treated as annual grasses but at least sorghum is harvested perennially in the tropics. If treated annually, the planting date for sorghum is July, whereas planting date can be March for perennially treated sorghum. Millets can be planted late May and early to mid-June (FAO 1995).

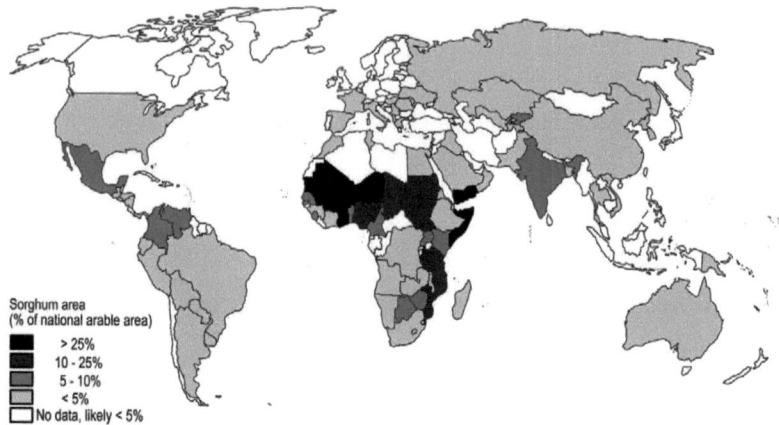

Figure 2: Relative importance of sorghum worldwide (modified from ICRISAT/FAO 1996)

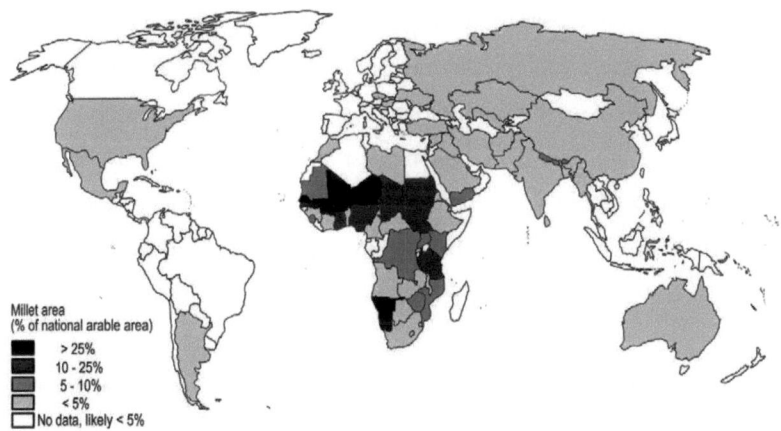

Figure 3: Relative importance of millets worldwide (modified from ICRISAT/FAO 1996).

In Northern, Western, Central and Southern Africa, the most important millet specie is pearl millet. About 70% of the pearl millet cultivation in Africa is grown in Western Africa. In Eastern Africa, but especially in the sub-humid uplands of Uganda, Kenya, Tanzania, Malawi, Democratic Republic of Congo, Zambia and Zimbabwe, the most widely cultivated millet specie is finger millet. It is less tolerant to droughts than pearl millet or sorghum, and therefore it has to be cultivated in savannah areas with at least moderate rainfall (Vietmeyer and Ruskin 1996). Table 3 shows the relative importance of millet species cultivated throughout the African continent.

Worldwide, sorghum and millets are cultivated on 40.9 and 34.8 million ha, respectively (wheat = 217.2 million ha). The estimated total production in 2010 was 55.7 and 31.6 million tons of sorghum and millets, respectively (wheat = 653.7 million tons). In 2010, the top 5 sorghum producing countries worldwide were United States, Mexico, India, Nigeria and Argentina. As mentioned before, the production in the United States and in the Latin American countries is almost completely for fodder production. Nigeria, Ethiopia, former Sudan, Burkina Faso and Niger were the top 5 sorghum producers in Africa using it as a food crop. Millet production in the year 2010 was highest in India followed by Nigeria, Niger, Mali and China. Beside the African countries mentioned before, Burkina Faso and Uganda belong to the top 5 millet producers in Africa (FAO 2009).

Table 3: Relative importance of millet species throughout Africa (Belton and Taylor 2002).

Region	Total millets (ktons)	Pearl millet (%)	Finger millet (%)	Teff (%)	Fonio (%)	Other millets (%)[1]
North Africa	554	98	2	0	0	0
West Africa	8986	95	0	0	4	1
Central Africa	447	87	13	0	0	0
East Africa	1547	35	50	9	0	6
South Africa	404	65	30	0	0	5
Africa	11938	76	19	2	1	3

[1] Other millets do not include proso or foxtail millets which are of local importance in Asia, South America, Australia and Europe but not in Africa.

1.4 Sorghum and millets as staple foods

1.4.1 Sorghum and millet consumption

Sorghum and millets are staple foods for large regions in the semi-arid tropics of Asia and Africa where they are the major sources of energy, protein, vitamins and minerals, especially for poor disadvantaged people. The highest daily sorghum consumption per capita in 2009 was reported in Burkina Faso followed by Eritrea, the former Sudan, Chad and Niger. Regarding millet consumption per capita per day, the consumption in 2009 was by far highest in Niger followed by Gambia, Mali, Burkina Faso and Chad (Figure 4) (FAO 2012).

In Niger, sorghum and millets together contributed to more than half of the daily energy intake in 2009. In Mali, the former Sudan and Chad, the contribution is around one third. The contribution exceeds one third in Burkina Faso and Eritrea (Figure 5) (FAO 2012). It has to be stressed that sorghum and millets are also processed into beverages, mostly alcoholic beers, which can contribute to daily calorie intake (Mwesigye and Okurut 1995). In some countries, such as Benin, Togo and India, consumption of sorghum and millets is extremely regional. In the North of Benin and Togo, sorghum and millet consumption is high (Kayode et al. 2005), but due to the high consumption of maize and rice in the South it does not appear relevant in the overall consumption statistics of the countries. In India, the overall daily per capita consumption of sorghum (rural + urban) was rather low with about 14 g. On the other hand, when looking at the sorghum-producing regions, such as the central inlands of India, the daily per capita consumption is extremely high with 206 g (Parthasarathy Rao et al. 2006).

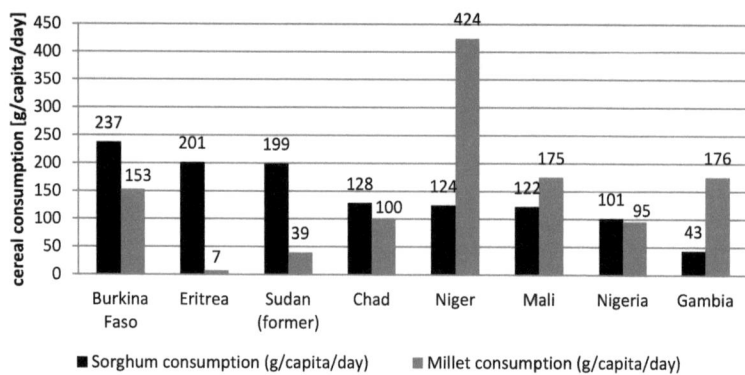

Figure 4: Eight highest daily consumption of sorghum and millets (g/capita/day) worldwide in 2009 (FAO 2009).

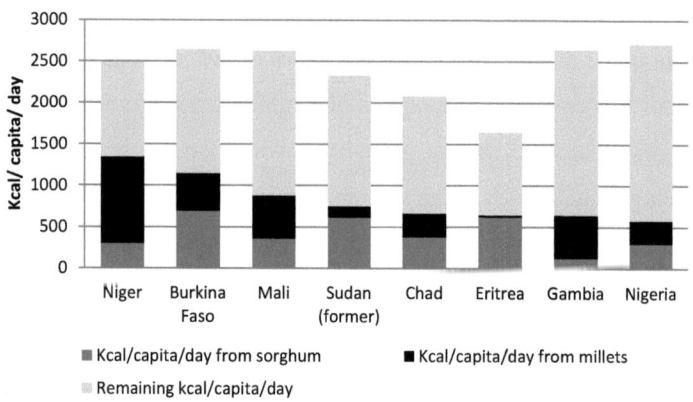

Figure 5: Daily energy intake from sorghum and millets (kcal/capita/day) and their contribution to total daily energy intake in the eight countries with the highest consumption of sorghum and millets in 2009 (FAO 2009).

1.4.2 Traditional sorghum and millet products

Sorghum and millets can be processed into various types of foods which can be classified in traditional products and non-traditional industrial products. In Africa and India, the traditional products, especially breads and porridges, play a much more important role than non-traditional industrial products (e.g. pasta or extruded products) (FAO 1995). In addition to breads and porridges, traditional products, such as steamed or boiled products, beverages and snack foods, were common. Table 4 summarizes the most well-known traditional products made of sorghum and millets frequently consumed in the semi-arid tropics. The table does not claim completeness. There are a vast number of foods and sometimes they have similar processing steps, but are named differently depending on the region or country (FAO 1995).

1.4.2.1 Fermented and unfermented bread made of sorghum and millets

Flat unfermented bread made of sorghum or millets is particularly popular in India. *Roti* and *chapatti* are the most common types of unfermented flat bread in India. Their preparation is very similar and starts with mixing water, flour and salt into a firm dough. Afterwards, the dough is divided into equal rounded portion which are then baked on a hot griddle. The two types of flat bread are mostly consumed with vegetable curry sauces or a traditional Indian sauce called *dal* (also spelled *dahl* or *daal*) which is mainly based on pulses (Olewnik et al. 1984; Murty and Subramanian 2009).

The most widespread fermented flat breads are: *dosa* (or *tosai*) in India (FAO 1995), *injera* (or *enjera*) in Ethiopia (Piccinin 2002), and *kisra* in Sudan (Badi et al. 1990). While the dough of *dosa* and *kisra* is usually fermented for 1 day or less (Rahman and Osman 2011), *injera* dough depending on taste habits can be fermented for 2 or 3 days (Blandino et al. 2003). *Injera* is a staple food only in Ethiopia and is preferably made of teff although sorghum, finger millet, maize and barley are also frequently used.

Sorghum flour is reported to be the second choice of flours preferred for *injera* preparation (Blandino et al. 2003). *Injera*, *dosa* and *kisra* are eaten at any time; breakfast, lunch, dinner, or just as a snack. They are main staple foods throughout all the socioeconomic classes.

1.4.2.2 Thick and thin porridges made of sorghum and millets

Porridges based on sorghum and millets are important foods in almost all sub-Saharan African countries and as well in India (Table 4). They are served with thick or thin consistency and are fermented or unfermented. Depending on the country, thick and thin porridges are known by different names but have similar preparation procedures (FAO 1995). Thick porridges are usually consumed with a sauce containing vegetables, meat or fish, oil, and/or spices (Rooney et al. 1986). Thin porridges (similar to creamy, free-flowing beverages) are mostly consumed as a breakfast and are frequently used as complementary foods for infants and young children (Onofiok and Nnanyelugo 1998). They are often sold early in the morning in a ready-to-eat form in front of the producer's home, on the street or in the local market (Tou et al. 2006; Mouquet-Rivier et al. 2008).

Food diaries, obtained from 13 rural households in Burkina Faso, showed that between 1.8 and 2.9 meals per day were a thick porridges called *tô*, made from pearl millet or red sorghum, and a sauce of leaves or groundnuts (Lykke et al. 2002). *Ugali* in Eastern Africa and *aceda* in Sudan are other popular thick porridges that can be made of sorghum and/or millets (FAO 1995). *Ben-saalga* is a very popular thin porridge frequently consumed as a breakfast in Burkina Faso. The preparation process involves soaking of cleaned whole grains of pearl millet, wet milling with water to obtain a smooth slurry, sieving the slurry to remove coarse particles, allowing the filtrate to sediment and ferment simultaneously overnight and decanting the supernatant soak water. After fermentation, a certain volume of decanted soak water is heated to boiling, and then some sedimented mass is gently stirred in to obtain the porridge of preferred thickness by gelatinization of the starch (Tou et al. 2006). Nutrient and energy density of *ben-saalga* is limited due to the low dry matter (DM) of the porridge; around 8–10 g/100 ml (Tou et al. 2007).

In Northern Benin, where sorghum represents an important staple crop, a number of sorghum foods are known. *Dibou*, a thick porridge, is the most important and it is consumed preferably at lunch or dinner together with a sauce. The preparation of *dibou* involves cleaning, milling, sieving, and cooking, but varies among regions or ethnic groups. Thin Beninese sorghum porridge similar to *ben-saalga* is called *koko* and it is often consumed with sugar, honey or roasted groundnuts. The preparation of *koko* usually involves 24 hours of fermentation followed by cooking. If fermentation is not applied, acidic taste is obtained by cooking sorghum slurry in acidic water, obtained from a previous fermentation (Kayode et al. 2005). *Ogi* is similar to *koko* and has several other names along the Western African costal region. It is considered as one of the most important complementary foods in Western Africa although it is also consumed by adults. *Ogi* is preferably made of maize, but sorghum and millets are also frequently used (Blandino et al. 2003). *Uji*, a thin porridge made from sorghum, finger millet or maize, is very popular in Eastern Africa and widely used as complementary food (Nout 2009).

Table 4: Common traditional foods made of sorghum and millets.[1]

Type of food/common name	Grains	Countries
Unfermented bread		
Roti/chapatti	Sorghum/millets[2]	India
Fermented bread		
Dosa	Sorghum/millets	India
Kisra	Sorghum/millets	Sudan
Injera	Sorghum/finger millet/teff	Ethiopia
Thick porridge		
Ugali	Sorghum/finger millet	Kenya, Tanzania, Uganda
Dibou	Sorghum	Northern Benin
Tô	Sorghum/millets	Burkina Faso, Mali, Niger
Aceda	Sorghum	Sudan
Thin porridge		
Uji	Sorghum/finger millet	Kenya, Tanzania, Uganda
Ogi/Koko	Sorghum/millets	Several West Africa countries
Ben-saalga	Pearl millet	Burkina Faso
Steamed/cooked products		
Couscous	Sorghum/pearl millet	North Africa, Sahel region
Tortillas	Sorghum	Central America
Alcoholic beverages		
Tchoukoutou/Dolo	Sorghum	Benin/Burkina Faso

[1] References for the different food types are in text.
[2] If not further specified, various millet species are used for the preparation.

1.4.2.3 Various sorghum and millet foods

Sorghum and millets are frequently used as raw materials to prepare fermented alcoholic and non-alcoholic beverages. Traditional opaque beers, such as *tchoukoutou* in Benin (Kayode et al. 2007a) or *dolo* in Burkina Faso (Pale et al. 2010), are popular beverages in several countries in Africa and as well in Asia (Nout 2009). Their preparation includes malting of sorghum or millets (soaking, germination and sun drying), brewing (mashing, boiling, filtration), and fermentation. Traditional opaque beers have a short-life of about one week, a low alcohol concentration (depends on the fermentation time), an acidic nature (pH ~3.2) and they contain considerable amounts of grain DM (FAO 1995; Kayode et al. 2007a). A few fermented non-alcoholic beverages made of millet (e.g. *Mangisi* in Zimbabwe) have regional importance (Blandino et al. 2003).

Couscous, a steamed agglomerated food, is the type of food made from sorghum and pearl millet which is also well-known in industrial countries. Couscous preparation involves milling, agglomeration by blending with water, and then cooking by steaming. It is particularly important in Northern Africa but also in countries of the Sahel region (Galiba et al. 1987; Anglani 1998). In some Central American countries, sorghum partially or sometimes even completely replaces maize in tortillas production (Choto et al. 1985). Sorghum and millet flours are also used in the local production of cookies and snack foods in developing countries (Anglani 1998).

1.5 Prospects for a sorghum and millet food processing industry

Sorghum and millets are regarded as non-profitable subsistence crops which respond weakly to fertilization (Sanders and Ouendeba 2012). This preconception goes hand in hand with the shift over the last 15 years towards the use of wheat, maize and rice at the expense of sorghum and millets in the average diets of the semi-arid tropics, especially those of urban consumers. However, sorghum and millets are still the most important food grains in Western Africa and they play a significant role in the cereal economy of many countries in the semi-arid tropics. Furthermore, it is likely that they become more important if cereals with higher drought resistance are needed due to climate change. Climate change is predicted to have severe effects on the water balance in the semi-arid and arid lands, as well as in medium or even high production zones, and hence will harm the agricultural sector in Africa (Fetene et al. 2011). Most such areas affected by climate change would be unsuitable for the production of grains other than sorghum and millets unless irrigation is available. Therefore, the establishment of a food processing industry comparable to wheat, maize or rice for sorghum and millets would be very much appropriate in developing countries (Ndjeunga and Nelson 1999). In Africa, the industrial use of sorghum and millets for food is almost entirely restricted to the commercial malting and brewing of the popular opaque beers in South Africa and Zimbabwe. Only in Eastern Africa, e.g. Kenya, a few complementary foods based on sorghum, pearl and finger millet have been commercialised (Obilana 2003). In India, the main industries using sorghum are the animal feed sector, alcohol distilleries, and starch industries (Kleih et al. 2007).

Only very little sorghum and millet enter the commercial market. In 1997, it was estimated that only about 8% of pearl millet production in Niger entered the market. The rest was consumed in the area in which it had been produced (Ndjeunga and Nelson 1999). The current sorghum and millet markets are characterized by high price variability and unreliable supply of high quality grain. There are many reasons for this. For example, lack in storage facilities often forces farmers to sell the grains at "collapsed" prices (Sanders and Ouendeba 2012). Improved technologies for grain cultivation, such as fertilizer application, soil restoration technologies and water conservations methods, are urgently needed for an increased and efficient supply in the future. Furthermore, the equipment currently used for threshing, dehulling and milling is antiquated and needs to be modified and adapted to meet industrial practices (e.g. central milling of grains) with lower processing costs. It has to be stressed, that new equipment and technologies are only efficient if the knowledge concerning the equipment and technologies is transferred effectively to the farmers. As sorghum and millets have been neglected in all research fields, chemical and physical characterization of sorghum and millets varieties for different food products is limited and needs more research. In terms of processed products, demand for these products has to be created to make it attractive for processors and markets need to be tested to gain information on costs and products preferred by customers. Formal and informal contractual schemes have to be established between buyers and producers of sorghum and millets. Lessons for the implementation of such schemes can be learnt from the current cotton or groundnut markets in developing countries (Ndjeunga and Nelson 1999).

2 Dietary Factors Affecting Iron Bioavailability from Sorghum and Millet Foods

The two main iron types present in food are haem iron and non-haem iron which follow different patterns and mechanisms for absorption (Lonnerdal 2010) (see also chapter 4.2: Interaction of malaria with human iron homeostasis). A third and a fourth type are ferritin iron and the so called contamination iron. Ferritin is a stable iron storage protein that can bind ~4500 atoms of iron. It is found in low concentrations in several types of plants. It is possible that iron from ferritin can be absorbed from intact or partially intact ferritin molecules, and thus be unaffected by dietary factors. However, studies on bioavailability and the mechanism of ferritin-iron absorption have reported conflicting results (Lukac et al. 2009). Contamination iron, originating from soil or from processing and cooking equipment, can be found substantially in the diets in developing countries (Harvey et al. 2000). Little is known about the mechanism of absorption of contamination iron, but it is assumed that it is rather poorly absorbable (Hallberg et al. 1983).

Haem iron in the diet originates from haemoglobin (Hb) and myoglobin in meat, poultry and fish, where it accounts for about 30–60% of total iron (Cook and Monsen 1976a). In the diet of industrialized countries approximately 10% of total dietary iron is haem iron (Carpenter and Mahoney 1992); nevertheless, it plays an important role because it is well absorbed at a relatively constant rate between 25–35% and provides about 40% of the absorbed iron (Monsen et al. 1978). The remaining iron in the diet (~90%) in the developed world is non-haem iron and its absorption is greatly affected by dietary enhancers and inhibitors. This is particularly important in developing countries where the iron intake of certain communities is almost completely based on non-haem iron (Carpenter and Mahoney 1992). The following chapters describe the main iron absorption enhancers and inhibitors. In contrast to the iron absorption inhibitors, none of the described iron absorption enhancers are native components of sorghum or millets. Therefore, iron absorption enhancers are only relevant in terms of composite meals based on sorghum and millets or as components in fortification premixes used in sorghum and millet products.

Similar to other cereals, the iron in sorghum and millets is mainly located in the germ (Abdelrahman et al. 1984; Favier 1989). Recently conducted studies reported very similar iron concentrations of 4.2, 4.1 and 4.7 mg iron/100 g sorghum, pearl and finger millet grains, respectively (Hama et al. 2012; Tripathi et al. 2012). The genetic variation, with varieties having an iron concentration up to 8 mg iron/100 g grains, seems to be higher in pearl millet than in sorghum (Reddy et al. 2005; Velu et al. 2007). Extremely high iron concentrations over 10 mg iron/100 g sorghum or millet grains, caused by exogenous iron contamination, have also been reported in the literature (Badau et al. 2005; Kayode et al. 2007b). Iron contamination can occur from soil, dust, metal parts or paint in threshing and milling equipment, rubber products (particularly silicon and neoprene), sample preparation, or grain handling (Pfeiffer and McClafferty 2007). Teff is one of the millet species highly susceptible to soil contamination during threshing due to the very small size of the grains (Besrat et al. 1980; Hallberg and Björn-Rasmussen 1981).

2.1 Iron absorption enhancers

Iron absorption enhancers described in the literature include ascorbic acid (AA), other organic acids and the so-called "meat factor". The potential enhancing effect of vitamin A, carotenoids, and non-digestible carbohydrates, such as inulin or oligofructose, remains unresolved because of the conflicting findings (Hurrell and Egli 2010). Iron absorption enhancers are not native compounds in sorghum and millets, and therefore only affect iron bioavailability from sorghum and millet foods if they are consumed as a part of a composite diet. This is most likely not the case in developing countries, where diets are often monotonous (Zimmermann et al. 2005a) and where availability of meat is limited for the majority of people (Schonfeldt and Gibson Hall 2012). However, AA can play an important role as a part of micronutrient premix when sorghum and millets foods are considered for iron fortification.

AA is the most potent enhancer of human non-haem iron absorption (Hallberg et al. 1986) counteracting the inhibitory effect of PA (Hallberg and Rossander 1984; Davidsson et al. 1994) and PPs (Siegenberg et al. 1991). There is evidence that the enhancing effect of AA is more pronounced in meals containing high amounts of PA than in low PA meals (Siegenberg et al. 1991). However, higher amounts of AA are needed to improve iron absorption in high PA meals (Davidsson et al. 1998) (see also chapter 3.3: Enhancing bioavailability of (bio)fortification iron). Two mechanisms are thought to be responsible for the enhancing effect of AA on iron absorption. Firstly, AA is able to reduce ferric to ferrous iron in the stomach and duodenum, and secondly, AA forms soluble complexes with ferric and ferrous iron at higher pH in the intestine. Both of these mechanisms enable iron to become available for being absorbed by enterocytes (Rossander-Hulthén and Hallberg 1996). It has been shown that the two mechanisms are related to the simultaneous consumption of AA and the iron containing meal (Teucher et al. 2004). The positive effect of AA on iron absorption has been shown to be dose dependent (Fidler et al. 2003b; Hurrell 2004a). Heat, light and oxygen sensitivity of AA have to be considered when adding it to foods in order to improve iron absorption. Processing and storage can significantly decrease AA concentration and consequently its enhancing effect on iron absorption (Hurrell 2004a).

Compared with ascorbic acid, the potential enhancing effect of organic acids on iron absorption has been investigated to a much smaller extent (Hurrell 2004a). In a study investigating the effect of adding 1 g of citric, L-malic or tartaric acid to a rice meal, iron absorption was increased 2- to 3-fold when adding the organic acids (Gillooly et al. 1983). The enhancing effect of citric acid was confirmed in a later study (Derman et al. 1987). A positive influence on iron absorption was also shown for lactic acid (Derman et al. 1980). The enhancing effect is most probably due to the chelating properties of organic acids increasing the solubility of iron (Teucher et al. 2004). However, results are not conclusive and for citric acid sometimes even contradictory (Hallberg and Rossander 1984). It is assumed that high concentrations of organic acids are needed to significantly improve iron absorption, and that these high concentrations will most likely lead to unfavourable organoleptic changes in most foods (Teucher et al. 2004). Therefore, adding these organic acids to iron-fortified foods has not been considered as a promising approach to increase iron bioavailability.

The enhancing effect of meat on human non-haem iron absorption is attributed to the so called "meat factor". Several human studies have shown an enhanced iron absorption by meat, but the mechanism and structures behind this effect are not yet identified (Björn-Rasmussen and Hallberg 1979; Hallberg and Rossander 1984; Boech et al. 2003). It has been proposed that meat peptides, such as cysteine-rich proteins, able to bind iron in the duodenum, protect the iron from binding to inhibitors and keep the iron in solution available for enterocyte absorption (Storcksdieck et al. 2007). Further research is needed to identify the exact structures responsible for the enhancing effect.

2.2 Iron absorption inhibitors

PA and PPs are native components in sorghum and millets and are the two major inhibitors of iron absorption. Therefore, the following two chapters are focused on these two components. Other dietary factors that have been identified as iron absorption inhibitors include calcium and different proteins. In relation to calcium, results are not consistent. While some studies have reported reduced iron absorption by calcium concentrations of 75–300 mg (Monsen and Cook 1976; Cook et al. 1991b; Hallberg et al. 1991), other studies found no significant decrease in absorption when 200–1000 mg calcium were added to a test meal (Troesch et al. 2009; Gaitan et al. 2011). In addition to non-haem iron absorption, calcium is reported to inhibit haem iron absorption at certain concentrations (Roughead et al. 2005; Gaitan et al. 2011). The mechanism behind the potential inhibition of non-haem and haem iron absorption has not been elucidated but could be related to the absorptive process itself or to the passage through the enterocytes (Thompson et al. 2010). Proteins reported to inhibit iron absorption include egg albumin (Hurrell et al.

1988), soy protein (Lynch et al. 1994) and milk proteins such as casein and whey (Cook and Monsen 1976a; Hurrell et al. 1989b).

2.2.1 Phytic acid

PA, *myo*-inositol-1,2,3,4,5,6- hexakiphosphat (IP6), is the most abundant form of phosphorus in seeds and to a slighter extent it is also accumulated in other plant tissues and organs such as pollen, roots, tubers and turions (Cosgrove 1980; Raboy 1997). In these tissues, phytate, the salt of PA, is the major storage form of phosphorus, inositol and minerals during germination and development (Raboy 2003). Other inositol phosphates, such as inositol tetraphosphates and pentaphosphates, are also present in plant tissue but to a far smaller extent (<15%) than phytate (60–80%) (Dorsch et al. 2003). In most cereals, phytate is found in the aleurone layer, pericarp and the germ (Odell et al. 1972), whereas in legumes, protein bodies of the endosperm or the cotyledon contain the highest proportion of phytate (Schlemmer et al. 2009).

2.2.1.1 Interaction of phytic acid with iron

As mentioned above, phytate plays an important role during plant ripening and maturation providing essential minerals to the growing seedling and to other plant compartments. Phytate is able to form strong, mainly insoluble complexes with divalent and monovalent minerals, such as iron, zinc, magnesium, copper, calcium and potassium, and therefore is a fundamental factor for good crop yields (Reddy 2002). Dietary phytate has been suggested to have beneficial effects such as protection against colon cancer, arteriosclerosis and coronary heart diseases (Kumar et al. 2010). On the other hand, the mineral binding properties of phytate are regarded as one of the causes for decreased zinc (Navert et al. 1985), calcium (Weaver et al. 1991), magnesium (Bohn et al. 2004), manganese (Davidsson et al. 1995) and iron (Hurrell et al. 1992) bioavailability from high phytate diets.

In developing countries, where the diet is mainly based on crops high in phytate, such as cereals and legumes, the daily intake of phytate is estimated to be much higher than in Western countries. Studies investigating phytate intake in developing countries, reported a daily PA intake up to 2000 mg and more (Harland et al. 1988; Ferguson et al. 1989; Khokhar et al. 1994; Murphy et al. 1995), whereas daily dietary phytate intake for typical Western style diets is assumed to be ~200–350 mg (Schlemmer et al. 2009). Dietary phytate intake per day also differs between urban and rural areas, between females and males, between young persons and elderly, and between omnivores and vegetarians (Schlemmer et al. 2009).

The high phytate intakes in developing countries most likely contribute to the low iron status in vulnerable population groups. Numerous iron isotope single meal studies have shown a strong inhibitory effect of PA on human iron absorption. Evidence of an inhibitory effect was firstly described in the early 1940s and has been repeatedly confirmed for cereals (Gillooly et al. 1984; Hurrell et al. 2003) and legumes (Davidsson et al. 2001a; Petry et al. 2010). The negatively charged PA forms strong insoluble complexes with the positively charged iron at neutral pH in the intestine and makes iron unavailable for absorption (Figure 6). There is evidence that iron absorption is not adapted during long-term PA consumption. A study comparing the effect of high phytate bran on iron absorption of vegetarian subjects with a high phytate intake over several years with a control group, showed no difference in iron absorption between the groups. Adding bran to the test meals decreased iron absorption in both groups by more than 90% (Brune et al. 1989a). The inhibitory effect of PA on iron absorption is dose dependent but not linear; 10 mg and 20 mg PA, added to wheat buns, decreased iron absorption by 20% and 40%, respectively, whereas 100 mg PA, added to wheat buns, decreased iron absorption by 60% (Hallberg et al. 1989a). Another study confirmed the dose dependent inhibition and showed that ascorbic acid can partly overcome the inhibitory effect of PA (Siegenberg et al. 1991).

The PA to iron molar ratio (PA:Fe) is suggested as indicator to assess inhibition of PA on iron absorption rather than total amount of PA. For cereal or legume meals containing no enhancers of iron absorption, a PA:Fe <1:1 is proposed to improve iron bioavailability. Preferably, the PA:Fe should be 0.4:1 in such meals or, if possible, PA should be removed completely (Hurrell 2004b). In composite meals with meat or vegetables containing some ascorbic acid, a PA:Fe <6:1 is seen as favourable (Tuntawiroon et al. 1990; Hurrell and Egli 2010). To decrease PA, mechanical processes, such as extraction, decortication, and milling, can be applied (Hurrell 2004b; Lestienne et al. 2007). Furthermore, enzymatic degradation of PA, which requires the activation of native phytases during soaking, germination, and fermentation (see following chapter) or the addition of exogenous phytases, is a promising approach to decrease PA (Hurrell 2004b) (see also chapter 3.3: Enhancing bioavailability of (bio)fortification iron).

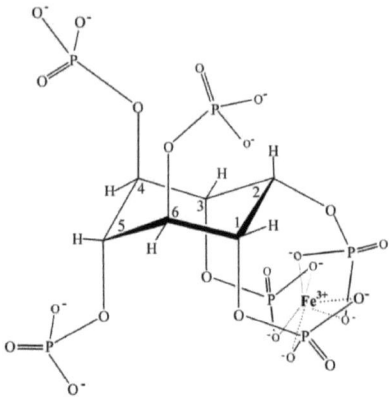

Figure 6: Molecular structure of mono ferric-phytate; iron (Fe3+-ion) forms a complex with phytic acid where all coordination sites of iron are occupied (Schlemmer et al. 2009).

2.2.1.2 Processing steps affecting phytic acid concentration in sorghum and millet foods

PA concentration in whole grain sorghum and pearl millet is reported to vary between 0.59–1.18% and 0.35–0.77%, respectively (Abdalla et al. 1998; Greiner and Konietzny 1999). The PA concentrations in some minor millets are as follows: 0.50% in finger millet, 0.67% in foxtail millet, 0.60% in common millet (Ravindran et al. 1994), 0.53–0.84% in teff (Abebe et al. 2007), and 0.41–0.67% in fonio (Koreissi et al. 2012). PA concentration in sorghum and millets is expected to be determined by environmental and genetic factors (Kayode et al. 2006).

Processing sorghum and millet grains into food can involve several steps that remove or degrade PA. One of these steps is decortication (also called dehulling or debranning) which is a widespread practice in sub-Saharan Africa to remove the bran from the starchy endosperm. In Burkina Faso, sorghum and millets used to prepare traditional *tô* paste are always decorticated in the case of sorghum and frequently decorticated in the case of millets. Decortication can be done manually by hand pounding with a wooden pestle and a mortar or mechanically using abrasion devices (Hama et al. 2011). Many studies have reported a significant decrease in PA after decortication of sorghum (Mahgoub and Elhag 1998; Hama et al. 2011), pearl millet (Pawar and Parlikar 1990; Sharma and Kapoor 1996; El Hag et al. 2002; Hama et al. 2011), or finger millet (Krishnan et al. 2012). The decrease in PA is due to the partial removal of bran and germ which contain the main proportions of PA in sorghum and millet grains (Simwemba et al. 1984). The magnitude of PA decrease depends on the decortication yield: the lower the decortication yield, the more PA is removed (Hama et al. 2011). It has to be stressed that decortication also removes iron, and therefore does not universally improve the provision of bioavailable iron (Lestienne et al. 2007; Hama et al. 2011).

Soaking of sorghum and millets is done as a pre-treatment to facilitate further processing steps such as wet milling, germination/malting or fermentation. It is generally done overnight, but can be as short as 15 minutes. During soaking, a small part of PA, which is water-soluble, leaches into the water. Additionally, native phytase is activated and hydrolyses phytate (Kumar et al. 2010). The magnitude of PA reduction is related to temperature, pH and soaking time (Greiner and Konietzny 1999). Soaking for at least 12 hours resulted in a moderate to high decrease of PA in sorghum and millets (Pawar and Parlikar 1990; Sharma and Kapoor 1996; Mahgoub and Elhag 1998; Sushma et al. 2008; Afify et al. 2011).

Germinated (or malted) millets are frequently used to prepare complementary foods in India (Hemanalini et al. 1980). In many African countries, malting of sorghum is a processing procedure traditionally used for the manufacture of local alcoholic drinks (Taylor and Dewar 2001). During germination, PA is degraded by native phytases. It has been shown that sorghum and millet have substantial phytase activity which increases during germination. However, compared to wheat this native phytase activity is much lower (Azeke et al. 2011). Several recent studies have demonstrated a vast reduction in PA after sorghum and millet germination (Mbithi-Mwikya et al. 2000; Makokha et al. 2002; Traore et al. 2004; Badau et al. 2005; Eyzaguirre et al. 2006; Abdelrahaman et al. 2007; Sushma et al. 2008; Azeke et al. 2011). In the study done by Badeau et al. 2005, the peak of PA degradation was reached after 72 hours. After 72 hours there was no further decrease in PA despite there was still about a tenth of the initial PA concentration. Traore et al. 2004 concluded that germinated red sorghum flour has favourable PA concentration to be used in complementary foods.

Fermentation of sorghum and millet flour is the processing step that has been most investigated and it is widely applied in traditional Indian and African foods (see chapter 1.4.2: Traditional sorghum and millet products). During fermentation, PA is degraded by native or microbial phytases from yeasts, moulds and/or lactic acid bacteria. Various *Lactobacillus plantarum* and *fermentum* strains expressing phytase have been found during the fermentation of *ben-saalga* pearl millet porridge (Songre-Ouattara et al. 2008). Compared to the other processing steps, fermentation seems to be the most effective approach for PA reduction (Mahgoub and Elhag 1998; Makokha et al. 2002). Depending on fermentation time, type of microorganisms involved and temperature, PA was almost completely degraded in some studies (Mahajan and Chauhan 1987; Khetarpaul and Chauhan 1990; Khetarpaul and Chauhan 1991). Fermentation of pearl millet seems to be particularly effective if a soaking step precedes fermentation (Sharma and Kapoor 1996). PA concentration in *ben-saalga*, the traditional millet-based Burkinabe porridge, decreased from 546 to 192 mg/100 g DM after fermentation (Tou et al. 2007).

2.2.2 Polyphenols

PPs are an extremely heterogeneous group of secondary plant metabolites naturally found in various plant species. Many of them are biosynthesized via the shikimate pathway which links the carbohydrate metabolism of the plant with the biosynthesis of aromatic compounds such as tyrosine, phenylalanine, tryptophan and PPs (Herrmann and Weaver 1999). PPs protect the plant against phytopathogenes, natural predators and UV radiation by absorbing light in the visible range (Friedman 1997). Furthermore, they play an important role for plant pollination by insects (Sisa et al. 2010). In the Western diet, PPs are abundant micronutrients that might play a role in the prevention of cancer and cardiovascular diseases (Manach et al. 2004).

2.2.2.1 Nomenclature and occurrence in sorghum and millets

In sorghum and millets, PPs are mainly located in the pericarp, but have also been detected in the testa, aleurone layer, and endosperm (Dykes and Rooney 2007). In general, PP concentrations in sorghum (0.40–1.35%) are substantially higher than in millets (0.14%), but a wide range of PP concentrations in sorghum and millets is reported in the literature (Dicko et al. 2002; Ragaee et al. 2006). One reason for this is the different PP concentrations among the different sorghum and millet varieties. Red and brown sorghums have higher average PP concentrations (1.35%) ranging from about 0.5–2.7% than brighter sorghum grains (0.40%) ranging from about 0.2–1.3% (Dicko et al. 2002). Another reason for the variation is the use of different analytical methods to measure the different phenolic structures (Brune et al. 1991). Methods measuring phenolic compound include the Prussian blue method (Price and Butler 1977), the Folin-Ciocalteu/Denis method (Burns 1963; Singleton et al. 1999), the ferric ammonium sulphate method (Beta et al. 1999) (which detect phenolics in terms of hydroxyl groups and report the amount as total phenolics), and the (modified) vanillin method (Burns 1971; Maxson et al. 1972) (which is specific for measuring phenolics with *meta*-oriented dihydroxyl phenolic groups). These methods use different extraction steps, different sample treatment and they express their results in equivalents based on different standards (tannic acid, catechin, gallic acid). This makes quantitative comparisons of PP concentrations very challenging (Earp et al. 1981).

From a chemical perspective, phenolic compounds can be defined as molecules with at least one aromatic ring bearing at least one hydroxyl moiety (Shahidi and Naczk 1995). Due to the heterogeneity of the different molecules, nomenclature of PPs is complex. They are usually classified into the following groups: 1) lignans and polymeric lignins, 2) stilbenes, 3) phenolic acids with the subgroup (hydroxy)benzoic acids and hydroxycinnamic acids and 4) flavonoids with subclasses (Figure 7) (Manach et al. 2004).

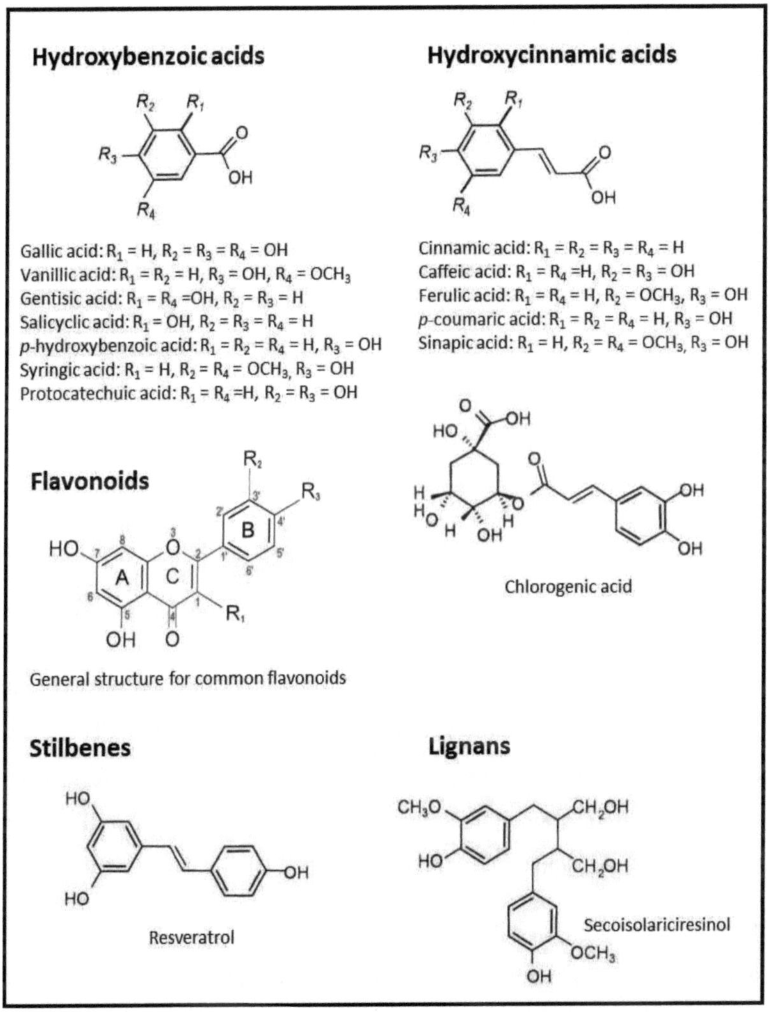

Hydroxybenzoic acids

Gallic acid: $R_1 = H, R_2 = R_3 = R_4 = OH$
Vanillic acid: $R_1 = R_2 = H, R_3 = OH, R_4 = OCH_3$
Gentisic acid: $R_1 = R_4 = OH, R_2 = R_3 = H$
Salicyclic acid: $R_1 = OH, R_2 = R_3 = R_4 = H$
p-hydroxybenzoic acid: $R_1 = R_2 = R_4 = H, R_3 = OH$
Syringic acid: $R_1 = H, R_2 = R_4 = OCH_3, R_3 = OH$
Protocatechuic acid: $R_1 = R_4 = H, R_2 = R_3 = OH$

Hydroxycinnamic acids

Cinnamic acid: $R_1 = R_2 = R_3 = R_4 = H$
Caffeic acid: $R_1 = R_4 = H, R_2 = R_3 = OH$
Ferulic acid: $R_1 = R_4 = H, R_2 = OCH_3, R_3 = OH$
p-coumaric acid: $R_1 = R_2 = R_4 = H, R_3 = OH$
Sinapic acid: $R_1 = H, R_2 = R_4 = OCH_3, R_3 = OH$

Flavonoids

General structure for common flavonoids

Chlorogenic acid

Stilbenes

Resveratrol

Lignans

Secoisolariciresinol

Figure 7: Chemical structures of polyphenols. Hydroxybenzoic and hydroxycinnamic acids are subgroups of phenolic acids. The listed hydroxybenzoic and hydroxycinnamic acids (except chlorogenic acid) are the most abundant phenolic acids in sorghum and millets (modified from Beecher 2003; Awika and Rooney 2004; Manach et al. 2004).

Lignans and polymeric lignins

Lignans consist of two phenylpropane units (Figure 7) and can occur as insoluble polymers with certain estrogenic characteristics. They are found in all higher plants with lignified tissue plants (Shahidi and Naczk 1995). The richest dietary source by far is linseed. Other dietary sources are legumes (lentils), certain cereals (wheat, rye and triticale), some vegetables (garlic, asparagus, and carrots) and some fruits (pears and prunes) (Thompson et al. 1991; Smeds et al. 2007). In sorghum, lignans have not been detected (Awika and Rooney 2004) and in millets only at very low concentrations (Penalvo et al. 2005; Smeds et al. 2007).

Stilbenes

Stilbenes are formed of two aromatic units which are linked by an ethane bridge (Figure 7). They can occur as monomers, polymers or glycosides. In the human diet, only low amounts of stilbenes are present. The mostly occurring stilbene is resveratrol which is found in grapes and wine (Fulda 2010). Resveratrol has been linked with anticarcinogenic effects. However, as only low quantities are present in the diet, its protective effect is not likely at normal nutritional intakes (Manach et al. 2004). Stilbenes have not been reported in sorghum and millets.

Phenolic acids

Two classes of phenolic acids are known: derivatives of benzoic acid (hydroxybenzoic acids) and derivatives of cinnamic acid (hydroxycinnamic acids) (Figure 7). Both types can occur in a free or bound form. In edible plants, the concentration of free hydroxybenzoic acids, such as gallic acid or 4-hydroxybenzic acid, is commonly low, with the exceptions of a few red fruits, black radish, onions, spices and herbs, and some tea leaves (Tomas-Barberan and Clifford 2000). Hydroxybenzoic acids are also part of complex structures, such as hydrolysable tannins and lignins. Hydrolysable tannins are either formed of esterified gallic acids units (gallotannin) or esterified hexahydroxydiphenic acid units (ellagitannins) (Haslam 2007). These complex structures composed of hydroxybenzoic acids are present in several edible plants of the two plant families *Leguminosae* and *Rosaceae* (e.g. beans, peas, lentils, nuts apple, pear, raspberry, plum, etc.). Hydroxybenzoic acids are also associated with proanthocyanidins/condensed tannins or, as in case of gallic acid, as a component of the proanthocyanidin structures (Haslam 2007; Serrano et al. 2009). Hydroxybenzoic acids are less common than hydroxycinnamic acids which primarily consist of caffeic, *p*-coumaric, ferulic, and sinapic acids. In processed food undergoing freezing, sterilization, or fermentation, these acids can be found in the free form, otherwise, they are found as glycosylated derivatives or esters of quinic acid, shikimic acid, and tartaric acid. Caffeic acid, both free and esterified, is commonly the most abundant phenolic acid. Chlorogenic acid, the ester of caffeic and quinic acid, is present in many fruits. Concentrations of up to 350 mg per single cup were reported in coffee (Clifford 1999).

Similarly to other cereals, ferulic acid is by far the most abundant phenolic acid in sorghum and millets (Sosulski et al. 1982; Hahn et al. 1983; McDonough and Rooney 2000). Ferulic acid is mainly present in the outer parts of the cereal grains, where up to 90% is esterified to arabinoxylans and hemicelluloses (Lempereur et al. 1997). Besides ferulic acid, protocatechuic, *p*-coumaric and cinnamic acid are the major phenolic acids in sorghum and millets (Hahn et al. 1983; McDonough and Rooney 2000). Other phenolic acids that have been reported in sorghum and millets include *p*-hydroxybenzoic, gentisic, salicylic (only in sorghum), vanillic, caffeic, sinapic acid (Figure 7). Gallic acid has also been detected in sorghum but only as a bound phenolic acid (Dykes and Rooney 2006). In sorghum, the phenolic acids are mostly present in bound forms (Hahn et al. 1983). Limited information is available on the proportion of free and bound phenolic acids in millets. A study investigating finger millet reported that most of the phenolic acids (71%) were present in the free form (Rao and Muralikrishna 2002).

Flavonoids

More than 8000 different flavonoids have been described so far and their numbers are still increasing. Flavonoids are characterized by two aromatic rings (A and B) which are connected together by a bridge of three carbon atoms that form an oxygenated heterocycle (ring C) (Figure 7). According to the binding of the B-ring to the C-ring, as well as the oxidation state and the functional groups of the C-ring, flavonoids are further divided into six subclasses: flavonols, flavones, isoflavones, flavanones, flavanols, anthocyanidins (Figure 8) (Manach et al. 2004). Flavonoids, except the flavanols, mostly occur as glycosides and other conjugates. The huge variety of different sugars forming glycosides with flavonoids contributes to the large number of different individual molecules that have been identified (Beecher 2003). Many different flavonoids have been reported in sorghum grains. Concentrations of flavonoids in sorghum are usually much higher than in other cereals and sometimes even exceed the concentrations in fruits and vegetables (Awika et al. 2004). The type and concentration of flavonoids differ considerably between sorghum varieties, and are controlled by a set of well documented genes (Rooney 2000). In millet grains, flavones are generally the only detectable flavonoids, but in some millets with pigmented pericarp and/or testa polymerized flavanols (condensed tannins) have been reported (Dykes and Rooney 2006).

Flavonols are usually present at relatively low concentrations of 15–30 mg/kg fresh food. However, they are the most ubiquitous flavonoids in foods and the major exponents are quercetin and kaempferol. Onions, curly kale, leeks, broccoli and blueberries are the richest sources. Other good sources are wine and tea which can contain up to 45 mg flavonols per litre (Manach et al. 2004). Flavonols usually occur as glycoside with glucose or rhamnose as the main sugar moieties, but also arabinose, galactose and xylose were reported as saccharide moieties (Bohm et al. 1998; Aherne and O'Brien 2002). Flavonols are relatively rare in sorghum and have not been reported in millets (Dykes and Rooney 2006; Awika 2011). Kaempferol 3-rutinoside-7-glucuronide, taxifolin and taxifolin 7-glucoside (Figure 8) are the only flavonols detected in red sorghum (Nip and Burns 1969; Gujer et al. 1986).

Flavones are much less common than flavonols. They are mainly reported in parsley and celery (Yao et al. 2004). However, nutritionally significant concentrations of flavones have also been reported in sorghum. Apigenin and luteolin (Figure 8), along with their 7-O-methylated derivatives, and some glycosides are the major flavones in sorghum (Dykes et al. 2009; Dykes et al. 2011). Accumulation of flavones in sorghum depends highly on genetic factors. Pigmented sorghum grains (white, lemon yellow, reddish/brown) with tan spike colours have much higher concentrations of flavones than pigmented sorghum (lemon yellow, reddish/brown, black) with red/purple spike colours (Awika 2011). Flavones (apigenin, luteolin, orientin, tricin, vitexin) have also been reported in pearl millet (Reichert 1979), barnyard millet (Watanabe 1999) and fonio. The concentrations of luteolin and apigenin (15 and 35 mg/100 g, respectively) in fonio were relatively high due to the large proportion of pericarp in the very small grains (Sartelet et al. 1996).

Flavanones are found at high concentrations in citrus fruits and are predominately glycosylated by a disaccharide; either a neohesperidose, which is responsible for a bitter taste, or a rutinose, which is tasteless. The most prevalent aglycones (not esterified with a saccharide) are naringenin in grapefruit, hesperetin in oranges, and eriodictyol in lemons (Manach et al. 2004). Relatively high concentrations of flavanones are accumulated by sorghum. The major flavanones detected in sorghum are primarily eriodictyol and naringenin (Figure 8) as well as their glycosides (Yasumats.K et al. 1965; Kambal and Batesmith 1976; Gujer et al. 1986). Similarly to other flavonoids, concentrations of flavanones in sorghum are highly influenced by genetics, particularly by genes coding for pericarp colour (grain colour). Sorghum with lemon yellow grains have the highest concentrations of flavanones which can be even higher than those reported in citrus fruits (Awika 2011; Dykes et al. 2011).

Isoflavones have a similar structure to estrogens, which confers them the ability to bind to estrogen receptors. This estrogenic activity is the reason why they are classified as phytoestrogens (Ward and Kuhnle 2010). Isoflavones are present almost exclusively in leguminous plants. In the human diet, soya and its processed foods are the main sources of isoflavones. The two main aglycones of isoflavones are genistein and daidzein (Figure 8), which commonly occur as their glycosides genistin and daidzin (Valls et al. 2009). Some isoflavones are sensitive to processing steps, such as heating and fermentation (Kudou et al. 1991). No reports have been found for isoflavones in sorghum and millets.

Flavanols (mainly flavan-3-ols, flavan-4-ols and their derivatives) are often polymerized into larger molecules called proanthocyanidins or condensed tannins (Escarpa and Gonzalez 2001). The term proanthocyanidins derives from their ability to form red coloured anthocyanidins when heated under acidic conditions (Cheynier 2005). The degree of polymerization varies from 2 to more than 10 monomers. The polymerization usually occurs via C4→C8 interflavan bonds, referred as B-type linkage. Additionally, ether bounds between C2→C7 can occur referred as A-type linkage (Figure 7). The most prominent flavanols are flavan-3-ols, such as (+)-catechin/(-)-epicatechin, (+)-gallocatechin/(-)-epigallocatechin and (+)-afzelechin/(-)-epiafzelechin, along with their oligo- and polymers procyanidins, prodelphinidins and propelargonidins, respectively (Serrano et al. 2009). In contrast to other flavonoids, flavanols normally only occur as aglycones (Escarpa and Gonzalez 2001). The highest concentrations are found in tea and cocoa followed by grapes and wine (Amarowicz et al. 2009; Serrano et al. 2009).

In sorghum, catechin and epicatechin are the major flavan-3-ols, and are mostly found as condensed tannins (Figure 8) (Awika et al. 2003b). Depending on the variety, the concentrations of condensed tannins in sorghum vary from below 0.10 to up to 1.97% (Dicko et al. 2002). Concentrations of flavan-4-ols, such as luteoferol and apiforol (Figure 8), derived from flavones, differ among sorghum varieties. Sorghums with black, brown or red pericarp have generally higher concentrations of flavan-4-ols than varieties with tanned pericarp colours (Dicko et al. 2005; Dykes et al. 2005).

Sorghum, some millet varieties and barley are the only cereals that accumulate condensed tannins in relevant amounts. Condensed tannins in sorghum are located in the testa, the thin layer sandwiched between the pericarp and the aleurone layer (Earp et al. 1981; Dykes et al. 2005). While the pericarp of sorghum, which defines the colour of the grain, is coloured by deoxyanthocyanidins without containing condensed tannins, the testa is only pigmented if condensed tannins were present. This implies that pericarp colour (= grain colour) is a reliable indicator for total PP concentration in sorghum but not for the concentration of condensed tannins (Boren and Waniska 1992; Dicko et al. 2002). Condensed tannins in sorghum are frequently of the B-type with (-)-epicatechin as extension units and catechin as terminal unit (Gupta and Haslam 1978; Gu et al. 2002; Awika et al. 2003b; Gu et al. 2003). However, a great diversity of condensed tannins has been reported over the last years, including tannins with A- and B-type linkages consisting of procyanidin (bearing catechin-groups) and prodelphinidin (bearing galloyl-groups) (Brandon et al. 1982; Krueger et al. 2003). Furthermore, proluteolinidin and proapigeninidin, two condensed tannins of flavan-4-ols (luteoferol and apiforol) often glucosylated and with flavanones as terminal units, were also detected in sorghum (Gujer et al. 1986; Beecher 2003; Krueger et al. 2003). Finger millet with brown or red pericarp is the only millet that has been reported to contain condensed tannins (Ramachandra et al. 1977). Most likely, brown teff and fonio varieties also contain condensed tannins but reports are lacking.

Anthocyanins are components of the epidermal tissue of flowers, vegetables, grains and fruits, to which they confer a pink, red, blue, or purple colour (Mazza 2007). Depending on pH, they can adopt different chemical forms, both coloured and uncoloured (Manach et al. 2004). Because the aglycone form (anthocyanidins) is highly unstable, they often occur as glycosides of pelargonidin, cyanidin, delphinidin, petunidin and malvidin (Mazza 2007). The most widespread anthocyanidin in food is cyanidin. Anthocyanins are mostly abundant in fruits, where they are found mainly in the skin. Other rich sources are red wine, a few cereal varieties, and certain leafy and root vegetables such as eggplant, cabbage, beans, onions and radishes (Clifford 2000). Anthocyanins are the major subclass of flavonoids studied in sorghum. However, quantitative data on these compounds is still rare (Awika and Rooney 2004). Unlike the common anthocyanins mentioned above, the anthocyanins in sorghum are unique since they do not contain the hydroxyl group in the 3-positition of the C-ring (Figure 8) and are thus called 3-deoxyanthocyanins. They usually occur as aglycones (Clifford 2000) and were reported to be more stable to change in pH than the anthocyanins commonly found in fruits, vegetables and other cereals (Awika et al. 2004). Analogously to the common anthocyanidins in fruits and vegetables, the 3- deoxyanthocyanidins concentrated in the bran are responsible for the pigmentation of sorghum grains. The orange luteolinidin, the yellow apigeninidin (Figure 8), as well as their 5-O- and 7-O-methyl derivatives, are the most abundant 3-deoxyanthocyanidins in sorghum (Dykes and Rooney 2006). In addition, cyanidin and pelargonidin have been reported in sorghum (Yasumats.K et al. 1965). Sorghums with black pericarp have the highest concentrations of 3-deoxyanthocyanidins followed by red and brown sorghum (Awika et al. 2004; Awika et al. 2005; Dykes et al. 2005).

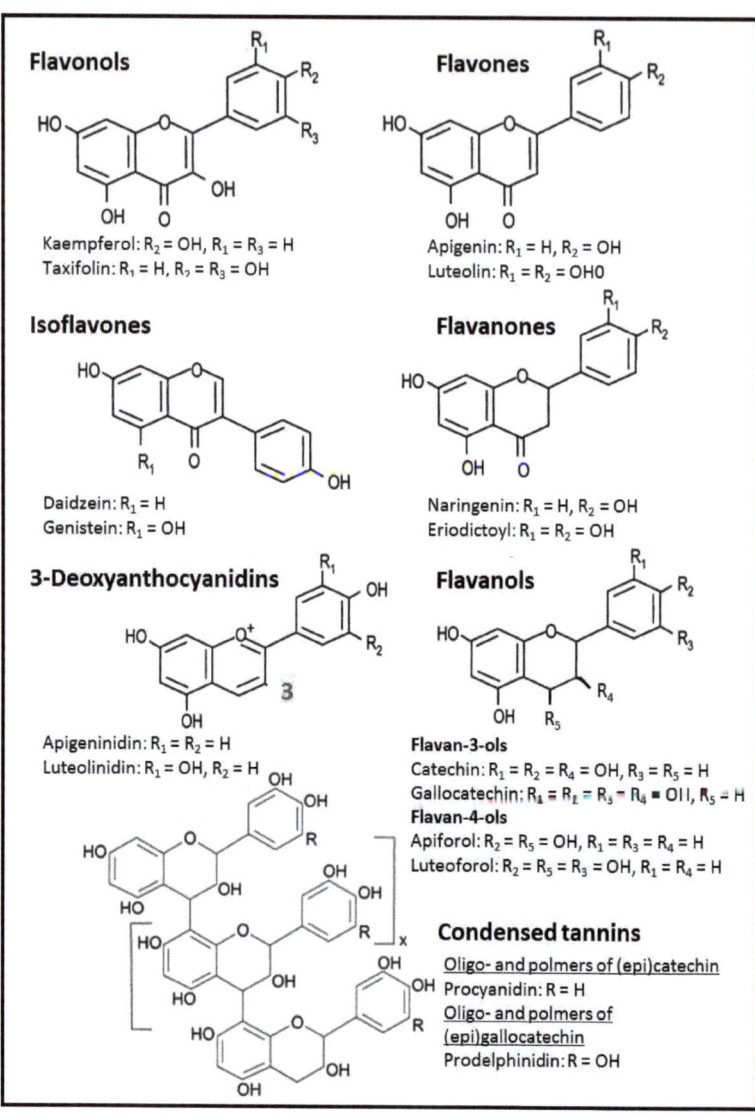

Figure 8: Chemical structure of flavonoids. Except for Isoflavones which are not present in sorghum and millets, the figure lists the most common flavonoids in sorghum and millets. Millets only contain flavones and rarely condensed tannins. Three-deoxyanthocyanidins in sorghum are a special appearance of the more common anthocyanidins (containing a hydroxyl-group at position 3), widely present in fruits and vegetables (modified from Manach et al. 2004).

2.2.2.2 Interaction of iron with polyphenols

The metal chelation property of some PPs is proposed as one of the mechanisms to explain the antioxidant role of PPs (Khokhar and Apenten 2003). In addition to metal chelation, some PPs can form complexes with proteins (Naczk et al. 2001) and polysaccharides (Barahona et al. 1997). The metal chelating ability of PPs is guaranteed by their structure and occurs by deprotonation of the hydroxyl-groups (Hider et al. 2001; Andjelkovic et al. 2006). In the literature, PPs with an *ortho*-dihydroxy (catechol) and/or trihydroxy-benzene group (galloyl), such as condensed and hydrolysable tannins, are considered to interfere with iron absorption by forming insoluble complexes in the gastro-intestinal lumen (Brune et al. 1989b; Hurrell et al. 1999). The iron-binding ability of catechol and galloyl groups was demonstrated using a spectrophotometric assay in which blue-coloured iron-trihydroxybenzene (galloyl groups) and green-coloured iron-dihydroxybenzene (catechol groups) were measured after the extraction of phenolic compounds from food (Brune et al. 1991). PPs with a single OH-group on the aromatic ring were found not to bind iron (Karamac 2009).

Complexes of iron and PPs are chelates, where PPs act as bi-dentate ligands binding iron at two sites. Some metal ions, such as iron, prefer an octahedral geometry, in which six atoms or groups of atoms are arranged around a central atom. Therefore, it is assumed that each iron atom preferably binds up to three catechol or galloyl groups, equivalent to a PP:iron binding ratio of 3:1 (Figure 9) (Perron and Brumaghim 2009). As mentioned above, deprotonation creating a highly charged oxygen centre is required for a strong complex formation. Deprotonation is favoured at high pH, but it has been shown that it can already occur at physiological pH between 5 and 8 (Hider et al. 2001; Purawatt et al. 2007). Complexes of ferrous (Fe^{2+}) and PPs are much less stable than complexes of ferric (Fe^{3+}) and PPs. It has to be stressed that several factors, such as pH, the ratio of metal ions to PPs in solution and the different PP structures, can influence and alter the above mentioned 3:1 binding ratio (Perron and Brumaghim 2009). It has been shown that in addition to the catechol or galloyl groups on ring B, flavonols, such as quercetin and myricetin, are able to bind iron between the oxygen at the 4-position and the hydroxyl group at the 3-position (ring C) or at the 5-position (ring A) (see Figure 7; general structure of flavonoids; R1 = OH) (Mira et al. 2002). However, these types of complexes have much less affinity to iron than the complexes formed by catechol and galloyl groups and if the phenolic compound is conjugated with a sugar moiety instead of having a OH-group at 3- or 5-position, the dissociable proton is not anymore available and it no longer binds to iron (Hider et al. 2001). In addition to flavonoids, also phenolic acids bearing catechol or galloyl groups (e.g. gallic acid, caffeic acid, protocatechuic acid) have iron-chelation properties (Andjelkovic et al. 2006).

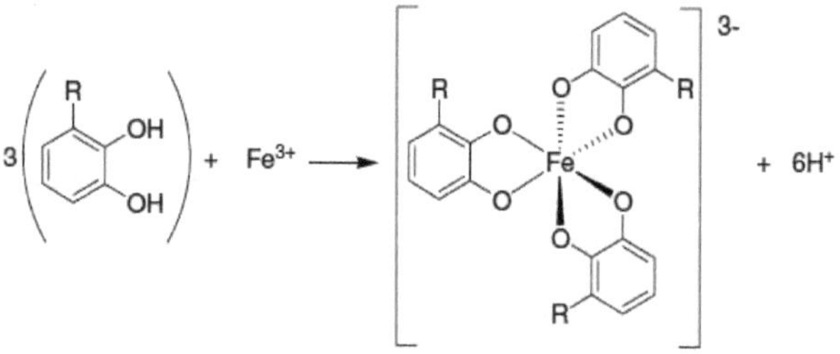

Figure 9: Assumed octahedral coordination geometry of general iron-polyphenol complexes. Deprotonation is required to coordinate the polyphenol ligands. R = OH = galloyls; R = H = catechols (Perron and Brumaghim 2009).

2.2.2.3 Human studies investigating the effect of polyphenols on iron absorption

The effect of PPs from different food sources on human iron absorption has been investigated in several studies. An inhibitory effect was first reported in a study with tea. Two-hundred ml tea prepared from 5 g dry tea, and served with different meals, decreased iron absorption by up to 80%. The negative impact of tea PPs even appeared in presence of ascorbic acid (Disler et al. 1975). In another study, the inhibitory effect of black tea, different herb teas and cocoa on iron absorption from a bread meal was compared with a control meal consumed with water. Beverages containing 100–400 mg total PPs/serving (gallic acid equivalents (GAEs); quantified by the Folin-Ciocalteau method) reduced iron absorption by 60–90%. At an identical concentration of total PPs, black tea was more inhibitory than cocoa, and herb teas (camomile, vervain, lime flower and pennyroyal) with peppermint tea as an exception, whose inhibition was equal to black tea (Hurrell et al. 1999). Red wine containing high amounts of proanthocyanidins and a total PP concentration of about 2–3 g/L has been reported to inhibit iron absorption when consumed with a simple bread meal. Compared with white wine with a total PP concentration of 0.19 mg/L, the iron absorption from the meal with red wine was 2–3 times lower; and 3–4 times lower than that from water served with the same bread meal. Iron absorption from the meal consumed with white wine was not significantly different to that from the meal consumed with water (Cook et al. 1995). Hurrell et al. 1999 compared these results with his own and concluded that black tea PPs are more inhibitory than PPs from herb teas, cocoa and wine, most likely because of the higher concentration of PPs with galloyl groups. In another study, red wine did not inhibit iron absorption when served with composite meal, whereas tea and coffee decreased iron absorption by 62% and 35%, respectively (Hallberg and Rossander 1982). Similarly, a further study reported a 50% reduction in iron absorption when a breakfast meal based on maize flour was consumed with 4 g espresso-type coffee or with 2 g tea (Layrisse et al. 2000).

The effect of PPs from different vegetables has been investigated in an early radioisotope study. In this study, 3 mg iron as ferrous sulphate ($FeSO_4$) was added to different vegetables. The results showed low iron bioavailability from aubergine, butter beans, spinach, brown lentils, beetroot greens and green lentils but moderate to good bioavailability from carrot, potato, beetroot, pumpkin, broccoli, tomato, cauliflower, cabbage, turnip and sauerkraut. The iron bioavailability from the different vegetables was strongly inversely correlated with total PP concentration in the vegetables (Gillooly et al. 1983). Another study looked at *yod kratin*, a leafy Thai vegetable containing considerable amounts of phenolic compounds. The study showed that a common portion size of *yod kratin* (20 g), equal to about 580 mg PPs (expressed as tannic acid equivalents), reduces iron absorption by 90% (Tuntawiroon et al. 1991).

PPs from oregano decrease iron absorption in a comparable way to beverages (Brune et al. 1989b), whereas the effect of rosemary is only moderate (Samman et al. 2001). A study investigating the effect of chili and turmeric, both sources of PPs, found an inhibitory effect of chili but not of turmeric. Twenty-five mg of chili PPs (GAEs; quantified by the Folin-Ciocalteau method) decreased iron absorption by 38%, whereas higher amounts of turmeric PPs did not inhibit iron absorption. The reason for this difference is not fully understood, but the authors argued that the reduced iron absorption from chili was due to its higher concentration of quercetin.

The authors finally concluded that iron bioavailability is not only affected by the quantity of PPs, but also and probably to a greater extent by their chemical structure (Tuntipopipat et al. 2006).

The effect of PPs on iron absorption from staple crops, such as sorghum and beans (common bean and cowpea), has been investigated in a few studies. A radioisotope study investigating the effect of sorghum tannins on iron absorption did not report a significant inhibitory effect, but a tendency towards increased iron bioavailability from low tannin sorghum (Radhakrishnan and Sivaprasad 1980). It has to be stressed that the tannin concentrations were only measured in the uncooked sorghum, and therefore the difference in tannin concentration in the cooked test meal was unknown. Gillooly and colleagues reported significant lower iron absorption from porridges prepared from a high-PP sorghum variety compared with porridges prepared from a low-PP sorghum variety. Both types of sorghum porridges contained relatively high amounts of PA (Gillooly et al. 1984). A more recent study showed that dephytinization did not increase iron absorption from high-tannin sorghum porridges, but increased iron absorption from low-tannin sorghum porridges by about 2-fold (Hurrell et al. 2003). High tannin sorghum varieties contain high amounts of catechol and galloyl groups (1350 and 550 mg/100 g grains, respectively) as parts of condensed tannins (Towo et al. 2003) which are expected to interfere with iron absorption.

PPs from common beans added as bean hulls to a bread meal decreased iron absorption by 14% when 50 mg PPs (GAEs; quantified by the Folin-Ciocalteau method) were added and by 45% when 200 mg PPs were added. No inhibitory effect was observed when 20 mg PP as bean hulls were added to the bread meal. Furthermore, the same study reported doubled iron absorption when PPs were removed from dephytinized bean porridge (Petry et al. 2010). A recent stable isotope study did not find a significant difference in iron absorption from white cowpeas (74 mg PPs expressed as GAEs/serving) compared with red cowpeas (158 mg PPs expressed as GAEs/serving). It has to be stressed that both cowpea varieties contained considerable amounts of PA (significantly lower in the white variety; 330 vs. 494 mg per serving) and that, depending on the PP structure, 74 mg PPs probably inhibit iron absorption to the same magnitude as 158 mg. The authors concluded that the high molar ratio of PA to iron seems to be the major determinant of iron absorption, rather than the difference in PP concentrations (Abizari et al. 2012a). A similar study with common beans reported a significantly lower absorption from high-PP bean meals (259 mg GAEs/serving) compared with low-PP bean meals (65 mg GAEs/serving). The PA concentrations were the same in both bean meals (~400 mg) (Petry et al. 2012). These results are inconsistent with the results from the cowpeas study. A possible explanation for this is the difference in PP concentrations between the meals, which was greater in the bean study, or the different PP structures between common beans and cowpeas. The study by Petry et al. 2012 also investigated iron absorption from a composite meal based on high-PP bean meals (176 mg GAEs/serving) and the same meal based on low-PP beans (38 mg GAEs/serving). The meals were fed as multiple meals over 5 days (10 portions) and no significant difference in iron absorption was observed. These findings suggest that the inhibitory effect of PPs on iron absorption may be less distinctive in a long-term perspective with composite meals containing considerable amounts of PA.

Brune and colleagues investigated the inhibitory effect of single PP compounds on iron absorption. They observed that adding 5, 25 or 100 mg tannic acid to a non-inhibitory test meal reduced iron absorption by 20, 67 or 88%, respectively. Tannic acid is commercially available condensed tannin, naturally not occurring in food. Similar inhibitory effects were found with naturally occurring PP compounds, namely gallic acid and chlorogenic acid, even though the latter was less inhibitory. Surprisingly, no inhibitory effect was observed when catechin was added to the test meals. Analogously to chlorogenic acid, the flavan-3-ol catechin has a catechol group which is expected to bind iron and inhibit its absorption. The reason why no effect was observed is not clear. The authors assumed that flavan-3-ols, such as catechin, are poorly water soluble, and therefore do not form complexes with iron in the intestinal lumen (Brune et al. 1989b). This assumption is contradictory with the studies that reported decreased iron absorption when tea was consumed with test meals. Tea contains large amounts of catechin or its derivatives, such as epigallocatechin gallate, epicatechin gallate, gallocatechin, epigallocatechin (Drynan et al. 2010), which either have galloyl or catechin groups or even both. Additionally, it has been shown that polymerised flavan-3-ols (= proanthocyanidins) are generally highly water soluble (Kawakami et al. 2010). One explanation for the discrepancy could be impaired solubility of pure catechin compounds compared with the solubility of catechins incorporated in the tea matrix. To date, there is still no clear evidence whether PPs with catechol or galloyl groups or both inhibit iron absorption from tea. Most of the foods or beverages contain a wide variety of different PPs which makes the prediction of interactions between PPs and iron very complex.

The effect of dietary factors, such as AA, or food additives, such as ethylenediaminetetraacetic (EDTA), on the interaction of PPs with iron has been investigated in some human studies, but more research has been done in *in vitro* studies. It has been shown *in vitro* that EDTA and AA are able to prevent iron chelation with tannic acid, but only EDTA can degrade already formed iron-PP-chelates and exchange the tannic acid (South and Miller 1998). Similar findings were made for EDTA and phenolic acids bearing catechol and galloyl groups (Andjelkovic et al. 2006). The studies were done with constant pH (4.4 and 7.4, respectively), and therefore findings cannot be directly applied to the human gastro-intestinal lumen with a pH gradient starting in the stomach (~1) and increasing until the duodenum (~6), where most of the iron is absorbed (Andrews 2008). In a radioisotope study, consumption of tea together with a Fe(III)EDTA-labelled test meal decreased iron absorption 7-fold suggesting that EDTA is not able to effectively prevent iron chelation with tea PPs (Macphail et al. 1981). In another study with tea, sodium iron EDTA (NaFeEDTA) partly overcame the inhibitory effect of PPs. Iron absorption from NaFeEDTA-labelled wheat bread rolls consumed with tea was significantly higher than absorption from the same meals labelled with $FeSO_4$, but still about 7-fold lower than NaFeEDTA-labelled meals without tea (Hurrell et al. 2000b). On the other hand, AA seems to have good potential to counteract the iron-binding of PPs in the human digestive tract. In a study using bread meals containing 420 mg tannic acid, already 30 mg of AA significantly improved iron absorption compared with control meals with no AA. The authors concluded that 50 mg AA is sufficient to overcome the inhibitory effect in meals containing >100 mg tannic acid (Siegenberg et al. 1991).

2.2.2.4 Improved iron bioavailability through reduced polyphenol concentration

In developing countries, where the diet is mainly based on crops high in PPs (e.g. sorghum or beans), the inhibitory effect of PPs on iron absorption may contribute to low iron status. However, so far no intervention study has investigated the influence of PPs on iron status and there is only some evidence from observational studies which reported that coffee intake is negatively associated with iron status (Fleming et al. 1998) or that tea PPs might have an adverse impact on iron status (Merhav et al. 1985; Gibson 1999). Nevertheless, absorption studies mentioned before (see chapter 2.2.2.3: Human studies investigating the effect of polyphenols on iron absorption) have clearly shown that some PPs strongly decrease iron bioavailability. To counteract the negative impact on iron bioavailability, the reduction of iron-binding PPs through plant breeding strategies or post-harvest procedures can be considered as appropriate approaches. However, as the targeted identification of iron-binding PPs is difficult and, as the prediction of which PPs will be affected by the intervention is challenging, most approaches target a general unspecific reduction of total PP concentration. When breeding plants with lower PP concentrations, the effect of the PP reduction on the plant has to be closely investigated before such an approach can be implemented. It would not be feasible if, for instance, the reduction in PPs compromises plant resistance against pathogens (Beebe et al. 2000).

PP concentration is affected by post-harvest processing such as washing, drying, milling, decortication, germination, fermentation, and thermal treatment (cooking, baking, extrusion, etc.). Depending on the processing step, PPs are either lost (e.g. during washing, soaking or decortication) or oxidized (e.g. during germination, fermentation or thermal treatment). Numerous studies have investigated changes of PP concentrations in sorghum and millets during post-harvest processing. Total PP concentration, expressed as tannic acid equivalents, decreased to one third of the initial PP concentration after soaking non-decorticated pearl millet for 15 hours in distilled water (Pawar and Parlikar 1990). Another study, which also looked into PP structures of sorghum and finger millet, found a significant decrease of total PPs and of PPs with resorcinol structure, but no significant change in PPs with catechol structure, after 24 hours water soaking (Towo et al. 2003). Depending on the decortication yield, total PP and tannin concentrations in sorghum and pearl millet significantly decreased after decortication (Sharma and Kapoor 1996; Lestienne et al. 2005; Dlamini et al. 2007). Germination can activate native polyphenol oxidases (PPOs) and has been reported to reduce tannin concentrations in light brown to dark brown finger millet. The reduction in tannin concentration increased over time and tannin concentration decreased by more than half at 72 hours (Rao and Deosthale 1988). Similar findings were reported for sorghum (Towo et al. 2003), pearl millet (Sharma and Kapoor 1996; Archana et al. 1998) and again finger millet (Mbithi-Mwikya et al. 2000; Towo et al. 2003).

Fermentation of pearl millet is known to decrease total PP concentration (El Hag et al. 2002; Elyas et al. 2002). Depending on the microbial strains involved in the fermentation, the reduction might be low (Khetarpaul and Chauhan 1991). Some studies even reported an increase in PPs (Khetarpaul and Chauhan 1990; Sharma and Kapoor 1996; Eyzaguirre et al. 2006). These differences can be explained by the activity of certain microbial PPOs which hydrolyse condensed tannins to lower molecular weight phenols resulting in an increase of total PP concentration (Sripriya et al. 1997). However, there is evidence that the low concentration of condensed tannins in pearl millet remains unchanged during fermentation, while total PP concentration decreases (Elyas et al. 2002). Studies on sorghum are more consistent reporting a significant decrease of total PPs during fermentation (Osman 2004; Towo et al. 2006; Dlamini et al. 2007; Rahman and Osman 2011). However, a study with red sorghum found an increase in certain phenolic acids and flavonoids after fermentation, while other phenolic acids and total PP concentration significantly decreased. The study did not investigate the changes in condensed tannins (Svensson et al. 2010). Thermal treatment, such as cooking (Mahgoub and Elhag 1998; Towo et al. 2003; Kayode et al. 2007b), baking (Awika et al. 2003a), or extrusion (Dlamini et al. 2007), were reported to significantly decrease total PP concentration in sorghum as well as in millets (Chowdhury and Punia 1997; Archana et al. 1998). The total PP concentration in *dibou*, the traditional thick Beninese porridge made of red sorghum, still contained about 80–160 mg PPs/100 g despite the reduction during cooking (Kayode et al. 2007b).

A relatively new approach to reduce PPs is the addition of exogenous PPO (e.g. catecholase, laccase, tyrosinase, cresolase). Catecholase is an enzyme that catalyses the oxidation of *ortho*-diphenols into quinones. Laccase oxidizes a wide range of phenolic substrates into quinones including *ortho*- and *para*-diphenols, and triphenols. Tyrosinase (animals/microorganisms) and cresolase (plants) are the same enzymes differentiated according to their origin. They preferably oxidize monophenols but are also involved in the oxidization of *ortho*-diphenols into quinones (Aniszewski et al. 2008). Incubation of red sorghum and red finger millet with tyrosinase over 16 hours resulted in a significant decrease in total PPs and in PPs bearing resorcinol or catechol groups. Using the dialyzability method, the *in vitro* iron accessibility from the two cereals was higher when they were incubated with phytase and tyrosinase than with phytase alone (Matuschek et al. 2001). In another *in vitro* study with sorghum porridges, the incubation with phytase and PPO increased the quantity of *in vitro* accessible iron. No increase was observed when porridges were incubated with PPO alone (Towo et al. 2006). The results of the two studies suggest that reduction of PPs by tyrosinase may have the potential to increase iron bioavailability but only if PA concentrations are low.

3 IMPROVING IRON NUTRITION THROUGH FORTIFICATION

Three conventional strategies are described to prevent iron deficiency (ID): 1) dietary diversification, 2) supplementation with pharmacological doses and 3) iron fortification of staple foods, infant formulas or condiments (Gibson 1997). Furthermore, in areas with chronic parasitic infections, the control of malaria and helminths infections is considered as necessary to successfully treat and prevent ID (Brooker et al. 2007).

Lack of dietary diversification is a major contributor to ID in poor disadvantaged people in the third world. The daily iron intake of these people relies on a monotonous diet based on cereals, legumes, roots and tubers with rather low iron concentration and low iron bioavailability. In contrast to wealthier people, poor people consume less nutrient-dense food such as animal products, fruits and vegetables. The aim of dietary diversification is to change dietary habits over a certain time by improving availability, access and utilization of foods with high iron concentration and high iron bioavailability (Gibson and Hotz 2001). Dietary diversification is regarded to be sustainable because it empowers individuals to take responsibility over the quality of their diet (Ruel and Levin 2001). However, in case of iron, possibilities to improve iron intake are relatively limited due to the high costs of animal foods and due to limited storage possibilities for such foods in developing countries (Gibson and Hotz 2001). Iron supplementation with relatively high doses of pharmaceutical iron is applied to targeted risk groups as capsules or tablets if substantial or immediate benefits are needed. It can be a cost-effective approach in specific situations, even though it is more expensive than other approaches such as food fortification (Baltussen et al. 2004).

Iron fortification is a useful approach to combat ID in cases where dietary diversification and supplementation are most likely to fail or not applicable. A recently published review including 60 trials concluded that the consumption of iron-fortified foods resulted in an improvement in Hb, serum ferritin (SF), and iron nutriture as well as in a reduced risk of remaining anaemic or iron deficient (Gera et al. 2012). The following chapters review the fortification approach as well as a promising novel approach called biofortification which is linked to traditional fortification and came into the focus of preventing ID over the last years (Nestel et al. 2006). The last two chapters review the current stage of sorghum and millet iron (bio)fortification and the methodologies to evaluate iron bioavailability from iron-(bio)fortified foods.

3.1 Conventional iron fortification

Conventional iron fortification refers to the addition of iron to a food, whether or not it is normally contained in the food, for the purpose of preventing or correcting a proven ID in a population or in a specific population group (WHO 2006b). It is considered as a safe and cost-effective approach for populations that consume significant quantities of industrially manufactured foods. Iron fortification originated in the 1940s in the United States and Europe when iron concentrations in low-extraction wheat flour were adjusted to the level of whole grains as a mean to combat ID. Nowadays, wheat flour is the food vehicle that is most often fortified with iron. Wheat flour fortification programs are in place or in the development stages in 78 countries (Ranum and Wesley 2008). In general, staple foods, such as cereal flours or condiments, are the most appropriate food vehicles for fortification although the choice of vehicle depends on the targeted population group (Hurrell et al. 2010).

Iron fortification can be roughly divided into three categories: 1) mass fortification, 2) targeted fortification, and 3) *in-home* fortification. The last two categories are closely linked to each other (WHO 2006b). Mass fortification is designed to increase the intake of bioavailable iron in the whole population, for the purpose of correcting ID in young children, adolescents, and menstruating women, without causing adverse effects in men and postmenopausal women, who most likely consume more iron than they require (Hurrell et al. 2010). In programs with targeted iron fortification, foods aimed at specific subgroups of the population are fortified, thus only the intake of these particular group is increased and not that of a whole population. Complementary foods for infants and young children, foods developed for school feeding programs, special biscuits for children and pregnant women, and rations for emergency feeding and displaced persons (e.g. blended foods managed by World Food Programme (WFP)) are examples for targeted food fortification. *In-home* fortification usually refers to a targeted fortification where the iron is added to foods at household level (e.g. complementary foods for young children). Examples for *in-home* fortificants include soluble or crushable iron tablets, iron containing micronutrient powders (Sprinkels^{TM}), and iron-enriched spreads (WHO 2006b). In cases where mass or targeted fortification is not possible, *in-home* fortification may be a useful approach to improve local foods for specific population groups such as infants and young children (Zlotkin et al. 2001). Examples of targeted and *in-home* fortification, and their efficacy, are discussed in more detail in the chapter 3.1.2: Iron-fortified complementary foods.

In addition to the targeted population and the food vehicle, the choice of iron compound is another key issue when planning iron fortification programs. The bioavailability of the iron compounds and its capability to cause undesirable organoleptic food changes differ among the different iron compounds and depend on the food vehicle. Iron fortification is technically challenging because the most bioavailable iron compounds tend to be those that cause the most adverse organoleptic changes (Hurrell 2002b). The different iron compounds are reviewed in detail in the following chapter 3.1.1: Iron compounds. Once the iron compound is chosen and dietary iron bioavailability (5%, 10%, or 15%) in the usual diet is estimated as well as information on iron and food intake in the targeted population is obtained, the level of iron fortification can be defined.

Iron intakes in population subgroups, such as children or menstruating women, are usually not normally distributed (Hurrell 2007). Therefore, WHO developed a full probability approach to define the iron fortification level in population subgroups with a certain iron intake. This approach uses tables to calculate the prevalence of inadequate iron intakes by different population subgroups in relation to the expected iron bioavailability in their regular diet (WHO 2006a). By knowing the inadequate intakes, fortification levels can be set based on the assumption of how much targeted food vehicle will be consumed by the target population and according to the relative bioavailability (RBV) of the iron compound used (Hurrell 2007).

The following chapter reviews the different iron compounds in relation to their bioavailability, their efficacy in improving iron status and their effects on organoleptic properties in various food vehicles. Regarding efficacy, only data from trials using a mass fortification approach (i.e. the fortified food vehicle is a staple food or condiment, potentially consumed by all population groups) are considered. The efficacy of targeted or *in-home* fortification, such as complementary foods or Sprinkles, is discussed later (see chapter 3.1.2: Iron-fortified complementary foods). Targeted fortification with infant formulas based on milk is not part of this review. Efficacy of iron-fortified infant formulas has been demonstrated in several trials (Moffatt et al. 1994; Walter et al. 1998).

3.1.1 Iron compounds

Several different iron compounds are currently used as food fortificants. These compounds can be broadly classified into four categories: 1) water soluble compounds, 2) iron chelates, 3) poorly water soluble compounds which are soluble in dilute acid, and 4) water insoluble compounds which are poorly soluble in dilute acid. The iron compound selected as fortificant for a specific food vehicle should have the highest RBV in the targeted food vehicle and, at the same time, not cause any unacceptable changes to the sensory properties of the food vehicle. RBV is a measure which scores the absorption of an iron compound by comparing its absorption to that of the reference iron compound $FeSO_4$, which is considered to have the highest absorption. Costs of the different iron compounds have to be considered when selecting a suitable compound (WHO 2006c).

3.1.1.1 Water-soluble compounds

Due to their high solubility in gastric juice, the water-soluble iron compounds have the highest RBV of the conventional available iron compounds. On the other side, these compounds often cause adverse effects on the sensory quality of foods, especially on colour and flavour.

$FeSO_4$ is the most commonly used water-soluble compound, because it has the maximum RBV (100%) of the conventional compounds and because it is the cheapest. The following section is focused on $FeSO_4$. Other examples for water-soluble compounds are: ferrous gluconate, ferrous lactate, ferric ammonium citrate, ferrous ammonium sulphate and ferric choline citrate. Information on their bioavailability is often limited to rats studies (Hurrell 2002a). Those with a similar solubility to $FeSO_4$ would be expected to have similar high bioavailability. In randomized controlled trials (RCTs), milk fortified with ferrous gluconate (RBV 100% in rat studies) was efficacious in reducing the prevalence of anaemia (Villalpando et al. 2006) and ID (Rivera et al. 2010) in young Mexican children (5.3 mg additional iron/day) when compared to a control group. On the other hand, no significant improvement in iron status was found in young Swedish children consuming milk fortified with ferrous gluconate (Virtanen et al. 2001). It has to be stressed that the concentrations of iron and AA were lower in the Swedish study than in the other two studies and that the Swedish children were fed *ad libitum*.

Ferrous sulphate

$FeSO_4$ has been widely used to fortify flours and is the recommended compound for low extraction wheat and degermed corn flour, pasta, and cereal-based complementary foods (WHO 2006c). However, rancidity and subsequent off-flavours have been reported in $FeSO_4$-fortifed wheat or other cereal flours stored for longer periods (Hurrell 1984; Hurrell et al. 1989a). In cocoa products, infant cereals, salt, and tortillas, $FeSO_4$ caused unacceptable colour changes (Hurrell 1984; Rao 1985; Hurrell 2002b). Furthermore, if multiple fortification of food is applied, free iron ions coming from the breakdown of $FeSO_4$ might oxidize some of the vitamins (WHO 2006c). Metallic taste was reported in bouillon cubes and fruit drinks fortified with $FeSO_4$ (Hurrell et al. 1989b) and it caused precipitation of peptides in soy and fish sauces (Fidler et al. 2003a).

Many studies have investigated the bioavailability of $FeSO_4$ in humans. A recent stable isotope study demonstrated that in contrast to insoluble iron compounds, absorption of $FeSO_4$ is up regulated in iron deficient women. The authors concluded that water-soluble iron compounds not only have better overall absorption, and therefore can be used at lower fortification levels, but they also have the added advantage that their absorption is up-regulated in ID (Zimmermann et al. 2011).

The efficacy of $FeSO_4$ has been shown in several RCTs using wheat flour or wheat flour biscuits (Zimmermann et al. 2005b; Sun et al. 2007; Biebinger et al. 2009), salt (Zimmermann et al. 2003) or fish sauce (Longfils et al. 2008) as food vehicles. The trials were done in school children, students or young women consuming 7.1 to 11.8 mg additional iron per day over 5 to 9 months. Iron status (SF plus Hb and/or serum transferrin receptor (TfR)) improved significantly in all trials. A study providing $FeSO_4$-fortified wheat flour (30 mg iron/kg) to anaemic Chinese students in a 6-months school feeding program reported significantly better Hb, SF and TfR concentrations compared to the control group (Huang et al. 2009). In a RCT in anaemic subjects, the prevalence of anaemia in Filipino children 6–9 year of age decreased from 100% to 38% after a 6-months intervention with fortified rice providing 10 mg additional iron per day (Angeles-Agdeppa et al. 2008). Furthermore, drinking water fortified with $FeSO_4$ and AA has been reported to increase Hb concentrations in children and adults, and SF concentration in adults (de Oliveira et al. 1996).

FeSO$_4$-fortified water improved Hb concentration in preschool children (Arcanjo et al. 2010).

3.1.1.2 Iron chelates

In the recent years some novel iron compounds, such as NaFeEDTA or ferrous bis-glycine chelate (FeBC), came into focus mainly because they provide better protection against iron absorption inhibitors than the other iron compounds (Hurrell 2002a; Bothwell and MacPhail 2004).

NaFeEDTA

The major advantage of NaFeEDTA is its chelating properties which protect iron from binding to iron absorption inhibitors. At the pH level of gastric juice, EDTA binds strongly to iron. Afterwards in the duodenum when pH rises, EDTA can exchange ferric iron for other metals. By doing this, EDTA prevents iron-binding to PA and perhaps PPs in the stomach and releases it for absorption in the duodenum (Hurrell 2002a). Human iron absorption studies reported that the iron absorption from NaFeEDTA-fortified whole grain wheat flour rolls was 4%, compared with only 1% from the same rolls fortified with FeSO$_4$, whereas in wheat rolls made from low extraction flour, the corresponding absorption values were 12% and 6%, respectively. The authors concluded that iron absorption from NaFeEDTA-fortified cereal foods high in PA is higher than when fortified with other compounds (Hurrell et al. 2000a). However, in meals with low or moderate phytate concentration, such as white rice and vegetables (Fidler et al. 2003a) or in composite meals including meat (Chang et al. 2012), the absorption from NaFeEDTA and FeSO$_4$ was similar. Regarding PPs, an *in vitro* study observed that EDTA can bind iron and prevent it from forming complexes with tannic acid at pH 4.4 (South and Miller 1998). In human studies, iron absorption from NaFeEDTA was significantly lower when NaFeEDTA-fortified test meals were consumed with PP-containing tea. However, iron absorption from NaFeEDTA-fortified meals was higher compared with the iron absorption from $FeSO_4$-fortified test meals consumed with the same tea indicating that, in contrast to $FeSO_4$, NaFeEDTA prevents at least some iron-binding by PPs (Macphail et al. 1981; Hurrell et al. 2000b).

NaFeEDTA usually causes fewer organoleptic changes than water-soluble iron compounds. It is highly useful in fermented sauces, such as fish or soy sauces, where most water-soluble iron compounds cause peptide precipitation during storage but not NaFeEDTA (Fidler et al. 2003a). In contrast to other iron compounds, such as $FeSO_4$, NaFeEDTA does not appear to promote fat oxidation reactions in stored cereals flours (Fidler et al. 2004b). On the downside, NaFeEDTA is about six to eight times more expensive than $FeSO_4$ and it has been shown that NaFeEDTA causes unwanted colour changes in corn and wheat flour, in tea and coffee after adding NaFeEDTA fortified sugar (Bothwell and MacPhail 2004), in cereals products with banana, and in chocolate drink powders (Hurrell 1997). Some of the colour changes are presumably due to the formation of complexes with PPs. Furthermore, NaFeEDTA can be degraded in some liquid products by ultraviolet rays from the sun. Storage of fortified fish sauce in clear glass bottles under daily sunlight resulted in a 35% loss of EDTA (Fidler et al. 2004b).

NaFeEDTA is approved by the Joint FAO/WHO Expert Committee on Food Additives (JECFA) and by the European Food Safety Authority (EFSA) for the use in supervised food fortification programs (JECFA 2007; European Food Safety Authority 2010). According to EFSA the use of NaFeEDTA as a source of iron in food is of no safety concern if it does not lead to an exposure to EDTA above of 1.9 mg EDTA/kg body weight/day (= 0.36 mg iron from NaFeEDTA/kg body weight/day). The acceptable daily intake (ADI) of 1.9 mg/kg body weight is particularly relevant when fortifying complementary foods for infants and young children in developing countries where prevalence of underweight is often high. A recently performed study concluded that if 2 mg iron as NaFeEDTA is provided to all infants 6–8 months of age, the percentage exceeding the ADI for EDTA would be <10% for populations with <30% of underweighted children. Therefore, 2 mg iron from NaFeEDTA can be used to fortify a daily serving of complementary food to ensure that EDTA levels are below ADI for this age group. Higher levels of iron fortification should be achieved by using additional other iron compounds than NaFeEDTA (Yang et al. 2011). Recently, EFSA approved the application of food-grade NaFeEDTA AkzoNobel (Ferrazone®) in food supplements, dietetic food and fortified foods but not in baby foods. The maximum amounts of NaFeEDTA (expressed as anhydrous EDTA) were set as follows: 1) 18 mg and 75 mg per daily dose for children and adults, respectively, in food supplements, and 2) 12 mg per 100 g final food in dietetic and fortified foods (European Union 2010).

WHO and FAO are recommending NaFeEDTA as an iron fortificant for the following food vehicles: high and low extraction wheat and corn flour, sugar, fish sauce, and soy sauce. For high extraction wheat or corn flour, NaFeEDTA should be preferred over other iron fortificants e.g. $FeSO_4$ and ferrous fumarate (FeFum) which can only supply sufficient iron if their amount is doubled or if they are added as encapsulated form (WHO 2006c).

Due to the strong chelating properties of EDTA, there has been some concern over the potential negative effect on the absorption of other minerals when EDTA binds to them in the lumen of the gut. However, an absorption study found no detrimental effect of NaFeEDTA on the metabolism of zinc, copper and calcium (Hurrell et al. 1994).

Quite a number of studies have tested efficacy of iron fortification with NaFeEDTA. Studies were done in staple foods and condiments mostly added to maize-based or rice-based composite meals. Except for one study using very low additional iron doses per day (1.3 mg) in South African children (van Stuijvenberg et al. 2008), all the other studies demonstrated improved iron status after the intervention. A study providing a relatively low dose of additional iron as NaFeEDTA in fortified maize porridge (3.5 mg iron) reported a significant reduction in ID and as well an improvement in SF and TfR concentrations when compared to a control group (Andang'o et al. 2007). The study done by Andang'o et al. also investigated a higher daily iron dose of 7 mg per day in the same maize porridge. After an intervention period of 5 months, all the iron status indicators (Hb, SF, TfR) improved in Kenyan school children and prevalence of anaemia, ID and IDA decreased significantly. Compared to the group receiving the low dose of NaFeEDTA, the positive effects on iron status and the ID reduction were much more significant in the group receiving the higher daily dose of NaFeEDTA. Two studies with fortified wheat flour providing 6 and 7 mg additional iron per day as NaFeEDTA, showed markedly improved iron status and reduced prevalence of ID in Indian school children and Chinese students, respectively (Sun et al. 2007; Muthayya et al. 2012). Furthermore, the study done by Sun et al. showed far better effect of NaFeEDTA in controlling ID and improving iron status than $FeSO_4$ or electrolytic iron. In another study with NaFeEDTA-fortified wheat flour (20 mg iron/kg) consumed during a school feeding program, iron status (Hb, SF, TfR) of anaemic students 11–18 years of age significantly improved compared to a placebo group (Huang et al. 2009). In a study with NaFeEDTA-fortified noodles providing 10.7 mg additional iron per day, dewormed Vietnamese school children improved iron status (Hb, SF but not TfR) compared to a dewormed control group. The beneficial effect of additional iron was more pronounced in a third study group receiving iron supplementation (56 mg iron as FeFum/day). However, there was no statistical difference between fortification with NaFeEDTA or supplementation with FeFum (Thi Le et al. 2006). In a recent RCT with cowpea, conducted in Ghanaian children 5–12 years of age from two rural communities, children in the intervention group received 10 mg additional iron as NaFeEDTA 3 days a week over a period of 7 months. At endpoint, iron status parameters (Hb, SF, TfR,) improved significantly and the prevalence of ID and IDA decreased significantly compared to the control group (Abizari et al. 2012b).

In Swiss women, who consumed an iron-fortified low-fat margarine with bread or pastries, providing 14 mg of additional daily iron as NaFeEDTA over a period of 32 weeks, body iron stores (BIS) significantly improved in the intervention group compared to the control group (Andersson et al. 2010). In a short-term study, fish sauce provided 10 mg additional iron as NaFeEDTA 6 days a week for 6 months. Iron status (Hb, SF, TfR) was improved, and prevalence of ID and IDA was reduced in Vietnamese women when compared to the control group (Thuy et al. 2003). Another short-term RCT reported increased Hb and SF concentrations after students 6–21 years of age consumed 114 meals seasoned with 10 ml fish sauce containing 10 mg iron as NaFeEDTA over a period of 21 weeks (Longfils et al. 2008). In a study with anemic school children, the low-NaFeEDTA group (consuming fortified soy sauce, providing 5 mg iron/day) and the high-NaFeEDTA group (consuming fortified soy sauce, providing 20 mg iron/day) showed increased Hb, SF, serum iron and transferrin (TF) concentration after 3 months. The two groups were not statistically different at endpoint (Huo et al. 2002).

Further studies with condiments were more long-term, lasting from 18 to 32 months. Three of them were classified as highly efficacious improving iron status and reducing prevalence of anaemia, ID or IDA. The studies were done with soy sauce providing 4.9 mg additional iron per day to a subsample of the whole Chinese population (Chen et al. 2005), with fish sauce providing 7.5 mg additional iron per day to women (Thuy et al. 2005), and with curry powder providing 7.1 mg additional iron per day to a subsample of the South African population (Ballot et al. 1989). A study with iron-fortified sugar in Guatemala delivering 4.6 mg additional iron as NaFeEDTA per day was only moderate efficacious improving only BIS in some subgroups of the targeted population (Viteri et al. 1995).

<u>Iron glycinate chelates</u>

The iron glycinate chelates include the following commercially available compounds: FeBC, ferric tris-glycine chelate (FeTC), ferric glycinate, and ferrous bis-glycinate hydrochloride. FeBC is the most investigated and most commonly used iron glycinate compound. It is used for the fortification of milk and dairy products in many Latin American countries, South Africa, and Italy (Hertrampf and Olivares 2004). In contrast to $FeSO_4$, it does not cause undesirable colour or flavour changes, and does not cause peroxidation of milk lipids, even after long-term storage (Allen

2002; Osman and Al-Othaimeen 2002). On the downside, FeBC promotes the oxidative rancidity in whole-maize meal much stronger than $FeSO_4$ and NaFeEDTA (Bovell-Benjamin et al. 1999), and it is one of the most expensive compounds for iron fortification at the moment (Allen 2002; Hertrampf and Olivares 2004).

There is evidence that a significant fraction of the iron in the FeBC is dissociated in the gastrointestinal tract, where it enters the exchangeable non-haem iron pool. On the other hand, Bovell-Benjamin et al. 2000 suggested that FeBC is absorbed as an intact chelate by an absorption pathway which differs from that of haem or non-haem iron. It is assumed that the iron-amino acid chelate protects iron from the inhibitors by being bound to glycine in a similar way than EDTA (Hertrampf and Olivares 2004). Two studies reported 2- and 4-fold higher iron absorptions from FeBC than from $FeSO_4$ when added to cereals foods containing PA (Bovell-Benjamin et al. 2000; Layrisse et al. 2000). In contrast, no difference in absorption between FeBC and $FeSO_4$ was found in another study from a high-PA whole grain complementary food (Fox et al. 1998). Additionally, in the study done by Layrisse et al., iron absorption from FeBC was increased after dephytinization and decreased by PPs from coffee and tea indicating that FeBC can only partly overcome inhibitors. The protection against inhibitors probably depends on the loss of chelated iron which mostly likely varies among different food matrixes (Hertrampf and Olivares 2004). FeTC, another iron glycinate chelate, has low solubility when pH is greater than 2 and is, therefore, less bioavailable than FeBC. In subjects with ID, iron absorption from both FeTC and FeBC was up-regulated (Bovell-Benjamin et al. 2000).

Foods fortified with FeBC have been evaluated in several intervention trials. An improvement in Hb concentrations and/or reduction in the prevalence of anaemia and/or ID were described in all of them. However, interpretation is difficult as most of the studies did not have a control group. Studies were done in children using fortified liquid milk (Iost et al. 1998; Osman and Al-Othaimeen 2002), dairy products (Fisberg et al. 1995; Miglioranza et al. 2003), wheat flour sweet rolls (Giorgini et al. 2001), beverages (Abrams et al. 2003; Ash et al. 2003) or bread (van Stuijvenberg et al. 2006). Trials using beverages or bread as the fortification vehicle had a control group and reported some improvement in iron status indicators. However, interpretation of the results in the study using beverages is challenging as they reported an increase in anaemia in both groups (although less in the intervention group). The study using bread showed increased Hb concentrations but no beneficial impact on SF concentrations. In a non-controlled study using sugar fortified with FeTC, Hb concentrations in children increased after a 6-month intervention (de Paula and Fisberg 2001).

3.1.1.3 Poorly water soluble compounds which are soluble in dilute acid

Iron compounds in this group are poorly water soluble but readily soluble in dilute acid. Depending on their solubility in gastric juice, they have a similar or slightly lower RBV than $FeSO_4$, but usually cause less organoleptic changes than water soluble compounds. However, it has been reported that FeFum causes unacceptable precipitation in soy sauce storage trials (Watanapaisantrakul et al. 2006). Presently, FeFum is used to fortify commercial infant cereals and ferric saccharate has been used in chocolate drink powders (Hurrell 2002a). Other compounds that can be considered for food fortification are ferrous succinate which seems to be highly bioavailable (Hurrell et al. 1989a) or ferrous citrate, ferrous tartrate, and ferric glycerophosphate which have moderately good bioavailability (Hurrell 2002a). More research is needed about these compounds. Currently FeFum is the only compound of this group which has been studied extensively.

Ferrous fumarate

FeFum, which is about 30% more costly than $FeSO_4$ (Hurrell 1999), is recommended for fortified foods for infants and young children (Hurrell 2010b). The recommendations are based on isotope studies showing similar iron absorption in adults for FeFum and $FeSO_4$ (Hurrell et al. 1989a; Hurrell et al. 1991; Hurrell et al. 2000b; Fidler et al. 2003b) and on the fact that FeFum has usually good sensory properties (Hurrell 2010b). A recently conducted study confirmed the comparable bioavailability of FeFum and $FeSO_4$ in non-anaemic women, young children and infants (Harrington et al. 2011). However, results are conflicting and earlier isotope studies in adults, iron-replete and also iron-deficient young children showed significantly higher absorption from $FeSO_4$ than from FeFum. Iron absorption from radiolabelled $FeSO_4$ was significantly higher than from FeFum in Chilean adults consuming a bread meal (Lopez de Romana et al. 2011). In Bangladeshi children with IDA, absorption from FeFum was only about 35% of that from $FeSO_4$ (Sarker et al. 2004) and Mexican children 2–3 years of age consuming a complementary food absorbed $FeSO_4$ about 3 times better than FeFum (Perez-Exposito et al. 2005). The reasons for these contradictory results are unclear, but potential explanations for the differences in children would be: the lower iron status in children resulting in greater iron absorption via up-regulation from $FeSO_4$ but not from

FeFum, a retarded dissolution of FeFum due to reduced gastric acid secretion in children; or an influence of food matrix on RBV (e.g. AA concentration) in some of the studies (Hurrell 2010b; Harrington et al. 2011). The latter is supported by a study showing that more soluble compounds, such as $FeSO_4$, are more sensitive to food matrix than less soluble compounds such as FeFum (Moretti et al. 2006b). In contrast to the assumption that absorption from FeFum is not up-regulated, a study with Ghanaian infants consuming FeFum-fortified porridge with 50 mg added AA reported higher absorption in the infants with IDA than in iron-sufficient infants or infants with ID (Tondeur et al. 2004). Other studies on the bioavailability of FeFum in infants and young children (Davidsson et al. 2000; Zlotkin et al. 2006) did not look at iron status and made no comparison with $FeSO_4$. Davidsson et al. 2000 compared the absorption from FeFum with ferric pyrophosphate (FePP), an insoluble iron compound, and found 2–3 timer higher absorption from FeFum.

Although results on iron absorption from FeFum in infants and young children are inconsistent, a RCT in South African infants reported a decrease in anaemia from 45% to 17% in the group consuming a multi-fortified maize meal (14 mg iron as FeFum/portion, AA:iron molar ratio (AA:Fe) = 1.6), whereas anaemia prevalence in the control group remained >40%. Furthermore, infants in the intervention group had significantly higher Hb and SF concentrations than infants in the control group (Faber et al. 2005). The relatively high iron concentrations may had contributed to the positive findings and may suggest that complementary foods for iron-deficient infants and young children should contain more iron in the form of fumarate as when the same food is fortified with $FeSO_4$ (Hurrell 2010b). In another study with rye bread providing 8.6 mg additional iron per day, Hb and SF concentrations in the intervention group were the same after 5 months as at baseline and were not different from the control group (Hansen et al. 2005).It has to be mentioned that the number of subjects (~20 women with marginal iron stores/group) was low in this study.

3.1.1.4 Water insoluble compounds which are poorly soluble in dilute acid

Relative to $FeSO_4$ these compounds are the least well absorbed iron fortificants (RBV = 20–75%) (WHO 2006c). During digestion, they dissolve slowly and incompletely in the gastric juice. The extent to which they dissolve depends on their physical characteristics (size, shape, and surface area of particles) and the composition of the meal. Physical characteristic can vary widely, and therefore prediction of absorption is challenging (Hurrell 2002a). Nevertheless, they have been widely used by the food industry because they rarely cause adverse organoleptic changes and because they are generally cheaper than the more soluble compounds. From a nutritional point of view, these compounds are regarded as the last choice for food fortification, especially for foods which are high in iron absorption inhibitors. If used for sensory reasons, it is recommended that they have an RBV of at least 50%, and that the amount of iron is twice as much as $FeSO_4$ so as to compensate for the reduced absorption (WHO 2006c). Insoluble compounds can be divided into two types: 1) iron phosphate compounds and 2) elemental iron powders.

Iron phosphate compounds

Phosphate compounds, such as regular FePP (mean particle size ~10–20 μm) and ferric orthophosphate (FeOP), are used to fortify rice, infant cereals, and chocolate drink powders (WHO 2006c). The RBV of FePP and FeOP depends on the mean particle size and has been reported to vary from 21–74% and 25–60%, respectively (Cook et al. 1973; Hallberg et al. 1989b; Hurrell et al. 1989a; Hurrell et al. 1991; Moretti et al. 2006b; WHO 2006c; Roe et al. 2009). A more novel phosphate compound, ferrous ammonium phosphate, has recently been tested in full-cream milk and showed better bioavailability than FePP, but was less bioavailable than $FeSO_4$ (Walczyk et al. 2012). Iron absorption from FePP in a soy-based meal compared with that from FeFum was significantly lower (1.3% vs. 4.1%) (Davidsson et al. 2000). Similar findings were made in a study comparing $FeSO_4$-absorption with the absorption from two FePP types with mean particle sizes of 8.5 and 6.7 μm, respectively (Fidler et al. 2004a). On the other hand, a study in adult women did not find a significant difference in iron absorption from infant cereals and yoghurt drinks fortified with dispersible micronized FePP (mean particle size = 0.3 μm) compared with the absorption from $FeSO_4$ (Fidler et al. 2004c). The small particle size in this study most likely increased the RBV of FePP. However, absorption from FePP with a slightly greater mean particle size of 0.8 μm was significantly lower than absorption from $FeSO_4$. The difference in absorption was greater in rice meals than in wheat-based meals and as well when AA was added. The authors concluded that not only the mean particle size strongly affects RBV of FePP but as well food matrix and iron status (Moretti et al. 2006b).

Numerous studies tested the efficacy of FePP in improving iron status. A RCT in India reported increased iron stores and decreased prevalence of ID in children who received rice fortified with 20 mg iron as micronized ground FePP when compared to the control group (Moretti et al. 2006a). In a RCT in anaemic Filipino children, FePP-fortified rice (10 mg additional iron/day) decreased the prevalence of anaemia to the same extent as $FeSO_4$-fortified rice providing the same amount of additional iron (Angeles-Agdeppa et al. 2008). In a study with a positive control group receiving $FeSO_4$-iron drops with rice (~72 mg additional iron/week), the increase in Hb and SF concentrations were higher in the group consuming FePP-fortified rice (~164 mg additional iron/week) (Beinner et al. 2010). In women, rice fortified with dispersible micronized FePP (13 mg additional iron/day) improved iron status (Hb, SF, TfR) compared to the control group after 6 months of intervention (Hotz et al. 2008). A recent RCT demonstrated increased BIS in Swiss women who received FePP- or NaFeEDTA-fortified margarine (14 mg additional iron/day) compared to women who received a placebo. BIS increased 2–3 times more in the group receiving NaFeEDTA fortification than in the group with FePP fortification (Andersson et al. 2010). Furthermore, salt fortified with micronized ground FePP, providing 18 and 18.6 mg additional iron over 10 months, significantly improved iron status in two RCTs in Moroccan school children (Zimmermann et al. 2004a; Zimmermann et al. 2004b). In Ivorian school children, the FePP-fortified salt was moderately efficacious when providing lower amounts of additional daily iron (10.5 mg) over a period of 6 months (Wegmuller et al. 2006).

Elemental iron powders

Elemental iron powders can be classified into five different types: 1) electrolytic, 2) hydrogen (H)-reduced, 3) carbon monoxide (CO)-reduced, 4) atomized (reduced), and 5) carbonyl iron powder. H-reduced and atomized iron powders are widely used to fortify wheat flour and electrolytic iron is frequently used to fortify cereal flours or breakfast and infant cereals. Depending on the manufacturing process, the different types of powder have different distinct particle size distribution, density, surface area, chemistry, and shape. All these factors influence the dissolution of the powders in the gastric juice, and therefore its bioavailability. It is even possible that different batches of the same type of powder have different dissolution characteristics (Hurrell 2002a).

Currently, electrolytic iron is the only iron powder recommended for food fortification. As it is only about half as well absorbed as $FeSO_4$, it should be added to provide double the amount of iron as $FeSO_4$ (Hurrell et al. 2002). Efficacy studies suggest that relatively large doses of at least 10 mg additional electrolytic iron per day can have a beneficial impact on iron status in women and children (Walter et al. 1993; Lartey et al. 1999; Zimmermann et al. 2005b; Sun et al. 2007). On the other hand, 20 mg additional iron per day as electrolytic iron, 4 times per week over 6 months, did not improve iron status in Ivorian school children with low prevalence of ID (Rohner et al. 2010). In another study, no changes in Hb were reported when women and children in Sri Lanka consumed 12.5 mg extra iron per day over a period of 2 years. SF has not been measured in this study (Nestel et al. 2004). Other efficacy studies using daily doses of electrolytic iron <10 mg did as well not report changes in iron status (van Stuijvenberg et al. 2006; Andang'o et al. 2007; van Stuijvenberg et al. 2008). In a study with anaemic Chinese students, the consumption of wheat flour fortified with electrolytic iron (60 mg iron/kg) in a school feeding program over 6 months decreased TfR concentrations but did not increase Hb and SF concentrations compared to the control group (Huang et al. 2009). CO-reduced and atomized iron powders are not recommended because their absorption is expected to be low and H-reduced or carbonyl powders require further evaluation (Swain et al. 2003; Lynch et al. 2007). So far, only one of five studies was able to show efficacy of H-reduced iron. The study was conducted in Thai women and a small reduction in ID was observed when 10 mg extra H-reduced iron per day was provided over a period of 9 months (Zimmermann et al. 2005b). All the other studies, sometimes using higher amounts of extra iron per day than the Thai study, did not demonstrate any benefit on iron status (Nestel et al. 2004; Winichagoon et al. 2006; Seal et al. 2008; Biebinger et al. 2009).

3.1.1.5 Encapsulated, micronized and nanostructured iron compounds

Research on iron fortification always looks for improved iron compounds that provide better bioavailability or better sensory properties than those currently available. In the recent years, the focus was on the following three approaches; 1) encapsulated, 2) micronized and 3) nanostructured compounds. These approaches are currently at different experimental stages. Encapsulated forms of $FeSO_4$ and FeFum are already commercially available and currently used in dry infant formulas and infant cereals in industrialized countries. Further application of encapsulated forms of iron compounds is likely in the future, but the increased costs of encapsulated compounds might limit their implementation. The purpose of encapsulation is the separation of the iron compound from the other food components, thereby mitigating undesirable organoleptic changes. Iron compounds can be encapsulated with hydrogenated vegetables oils, maltodextrin, mono- and diglycerides and ethyl cellulose (WHO 2006c). Coating materials have been reported to decrease the RBV of iron compounds in rat assays (Hurrell 1985) and also in a human study using encapsulated FeFum (Liyanage and Zlotkin 2002). The decrease in RBV of the encapsulated compound depends on the capsule:substrate ratio (Wegmuller et al. 2004; Zimmermann 2004). Despite the lower RBV of encapsulated compounds, three human studies showed efficacy of encapsulated $FeSO_4$ when added to salt and wheat-based biscuits (Zimmermann et al. 2003; Zimmermann et al. 2004a; Biebinger et al. 2009).

Insoluble irons salts, such as FePP, can be micronized by physical (mean particle size: 2–3 µm) and chemical (mean particle size: <1 µm) processes and thereby achieve better RBV in humans (Fidler et al. 2004c; Wegmuller et al. 2004). Efficacy of foods fortified with micronized FePP has been shown in several studies (Hurrell et al. 2010) (see also chapter 3.1.1.4: Water insoluble compounds which are poorly soluble in dilute acid). On the downside, smaller particle sizes can cause unfavourable organoleptic changes in fortified foods (Fidler et al. 2004c).

Still in a development phase are nanostructured iron compounds with a mean particle size of ~11 nm. In studies with rats, nanostructured powders of insoluble iron compounds had comparable RBV to $FeSO_4$ (Rohner et al. 2007; Hilty et al. 2010). Furthermore, the iron nanoparticles caused less colour changes when added to dairy products containing chocolate or fruit compared with the changes caused by $FeSO_4$ and FeFum in the same foods (Hilty et al. 2009). On the downside, production costs are high and the use of nanoparticles is linked with toxicity concerns. Acute toxicity of orally ingested copper and zinc nanoparticles has been demonstrated in animal models (Chen et al. 2006; Wang et al. 2006). Moreover, nanoparticles are possibly taken up by persorption and/or absorbed by gut-associated lymphoid tissue and pass through the mesenteric lymphatics to the venous circulation (Jani et al. 1990). This hasn't been shown for iron nanoparticles. No stainable iron was detected in the gut wall, gut associated lymphatics or other tissue after feeding rats with nanostructured iron compounds (Hilty et al. 2010). In conclusion, nanosized iron compound are a promising novel approach to control sensory changes in iron food fortification but more research on toxicity and bioavailability is needed (Hilty et al. 2009).

3.1.2 Iron-fortified complementary foods for developing countries

Complementary foods are introduced at around 6 months of age when human milk can no longer provide all the energy, protein, and micronutrients necessary to meet the infant's requirements for growth and development. They are the first solid foods eaten by the infants and they usually are part of a mixed diet that still includes breastfeeding. Because of the extremely low iron concentration in breast milk (Domellof et al. 2001), the provision of sufficient iron through complementary foods is particularly important. Therefore, the use of fortified complementary foods to prevent ID is recommended in several guidelines for complementary feeding (WHO 1998, WHO 2001).

Complementary fortified foods for developing countries can be divided into fortified blended foods (FBFs) and complementary food supplements such as lipid-based nutrient supplements (LNSs) or micronutrient powders (MNPs) (de Pee and Bloem 2009). There are other types of fortified complementary foods such as cookies or compressed bars. These types of fortified complementary foods have several disadvantages such as the risk of choking in younger children, the promotion of dental caries and they do not encourage an active and responsive feeding by the mother (Dewey et al. 2009a). The following three chapters are focused on

FBFs, MNPs and LNSs, which are reviewed in relation to their efficacy in improving iron status of infants and young children. Manufactured complementary foods commonly used in industrialized countries, such as infant formulas, are too expensive for the poor in the developing world and are not included in this review.

3.1.2.1 Fortified blended foods

FBFs are commercially produced cereals based on rice, wheat, corn, oat, or millet; soy, peanuts, or milk; sugar; and oil. They are used as a replacement for the traditional local porridge or in addition to traditional porridge used in complementary feeding of infants and young children (Dewey et al. 2009a). Many of the FBFs, such as the corn-soy blends and wheat-soy blends which are the products of choice to be provided to moderately malnourished children as well as to other vulnerable groups (pregnant and lactating women and people chronically ill with HIV/AIDS or tuberculosis) in food-assistance programs run by WFP, United Nations Children's Fund (UNICEF), and the US Agency for International Development (USAID), contain only cereals (75–80% corn or wheat), legumes (20–25% soybeans), and a micronutrient premix adapted to the targeted population group. Due to the protein concentration and quality of soybeans (total protein digestibility-corrected amino acid score (PDCAAS) of corn-soy blend: 65%) and the additional micronutrients, FBFs were regarded as being of reasonably good nutritional value for moderate costs.

Small-scale production of FBFs for the local market is implemented in several developing countries (de Pee and Bloem 2009). An example of such a small-scale production is *Misola*, a multi-fortified FBF based on millet, soy, and peanut, locally manufactured in several Western African countries (Simpore et al. 2006). Most of the FBFs are only partially adapted to young moderately malnourished children and have several disadvantages such as: 1) inadequate amount of some required nutrients, 2) large amount of antinutrients (e.g. PA) and fibers, especially when prepared from non-dehulled soybeans and non-degermed, non-dehulled maize or wheat, 3) low energy density per serving (bulky foods), 4) low overall energy (fat) concentration and low concentrations of essential fatty acids, 5) limited availability in rural areas, and 6) when centrally processed, they are still not affordable for the most poor people despite the moderate costs (Adu-Afarwuah et al. 2008; de Pee and Bloem 2009; Michaelsen et al. 2009). The issue of lack in energy per serving has been partly addressed by providing more sophisticated FBFs with oil and sugar (de Pee and Bloem 2009). Furthermore, WFP has developed an improved corn-soy blend with milk. There is some evidence that milk plays an important role in linear growth of young malnourished children (Hoppe et al. 2008). However, milk or milk powder is not locally available in developing countries. Refining cereals and legumes can significantly reduce PA concentrations in FBFs. However, low iron bioavailability due to PA remains a crucial issue and options to reduce its concentration need to be further explored. In a study that reviewed iron and PA concentrations in plant-based complementary foods used in low-income countries, only 27% of indigenous and 22% of processed complementary foods had a PA:Fe less than 1, although 89% of the processed complementary foods claimed to be fortified specifically with iron (Gibson et al. 2010). Exogenous phytase added to processed dry products or to prepared products (*in-home* fortification) is a novel promising approach to degrade PA in FBFs. It has to be taken into account that the heat-sensitivity of phytase limits its application before or during processing (de Pee and Bloem 2009) (see also chapter 3.3.3: Reduction of dietary phytic acid).

Efficacy of FBFs has been shown in a non-randomized controlled study in Ecuadorian infants using a commercially available product called *Mi Papilla* (10 mg $FeSO_4$/100 g product). The daily ration of *Mi Papilla* was 65 g providing 6.5 g additional iron per day over a period of 11 months. At endpoint, the children in the intervention group had significantly higher Hb concentrations, lower prevalence of anaemia and as well higher weights. No difference between the groups was observed in SF and TfR concentrations (Lutter et al. 2008). *Nutrisano*, a complementary food based on milk, hydrolysable corn starch, and a micronutrient premix, providing 10 mg additional iron per day either as $FeSO_4$ or ferrous gluconate during 6 months, did not increase Hb and SF concentrations of toddlers 12–30 months old when compared to a control group. However, TfR concentrations were lower in the intervention groups than in the control group and TfR and SF concentrations in the intervention groups were improved at endpoint compared with baseline (Shamah-Levy et al. 2008). A recent study in young Bangladeshi children showed that FeFum (9.3 mg) is as useful as FePP and $FeSO_4$ to maintain Hb concentrations >105 g/L and to prevent ID when feeding a FeFum-fortified infant cereal (*Cérélac*) over 9 months. No control group was included in this study (Davidsson et al. 2009). Pakistani infants who received a FePP-fortified wheat-milk complementary food (7.5 mg iron/100 g) over 12 months had significantly better iron status (Hb, SF) than infants in the control group, but did not differ from infants in a group receiving the same amount of iron as FeFum over the same time (Javaid et al. 1991). Approaches using FBFs can only be successful and sustainable when a nutrition education network by local authorities establishes increased awareness among families about the fortified complementary food and the optimal feeding practices (Bruyeron et al. 2010).

3.1.2.2 Micronutrient powders

In-home fortification of foods with MNPs is currently recommended by WHO to improve iron status and reduce anaemia among infants and children 6–23 months of age (WHO 2011a). The idea of MNPs (also known as Sprinkles^{TM}) was formulated in 1996 when a group of consultants determined that the available interventions, such as syrups and drops, were not effective for the prevention of childhood IDA (Nestel and Alnwick 1996). Sprinkles and "sprinkles-like" products are sachets containing a mixture of micronutrients in powder form which is easily sprinkled onto foods prepared at home (Schauer and Zlotkin 2003). They are optimally poured into semi-liquid foods after the food has been cooked and is at a temperature acceptable to eat. Sprinkles should be added to an amount of food that the child can consume at a single meal and the child should not share the food as one single package of Sprinkles is adapted to its daily micronutrient requirements (Sprinkles Global Health Initiative 2012). Based on several studies among children under five, MNPs are well accepted (Zlotkin et al. 2001; Zlotkin et al. 2006; Ip et al. 2009; Tripp et al. 2011) and are also a promising way to address widespread deficiencies of other micronutrients (de Pee et al. 2008). Formulations of MNPs can be modified to meet micronutrient requirements of other population groups such as pregnant women (WHO 2011b).

Efficacy of MNPs in children to prevent and treat ID or IDA has been tested in various setting over the past years (De-Regil et al. 2011). Many of the earlier studies used high daily iron doses between 30–80 mg (Zlotkin et al. 2003; Christofides et al. 2006; Sharieff et al. 2006) for which the term fortification seems to be inappropriate. At that time, Sprinkles were defined as hybrid between targeted fortification and supplementation (Christofides et al. 2005). Once studies showed that MNPs with lower iron doses had the same effect on Hb concentrations as high doses (Christofides et al. 2006; Hirve et al. 2007), most of the following studies changed to iron doses of 12.5 mg or even lower. RCTs in infants or young children using 12.5 mg iron doses daily or every 2 days were effective in preventing and treat anaemia in a 12-months intervention (Giovannini et al. 2006) or found a positive effect on Hb concentration in 2-months trials (Menon et al. 2007; Lemaire et al. 2011), as well as on SF and TfR concentrations when administered during 6 months (Adu-Afarwuah et al. 2008). In a recent study with Cambodian infants, daily micronutrient Sprinkles during a 12-months period significantly reduced anaemia and ID compared to a control group (Jack et al. 2012). $FeSO_4$-powder sprinkled onto rice porridge (10 mg additional iron/day) consumed for 6 months by North Korean infants has been reported to increase Hb and SF concentrations and to reduce anaemia and ID when compared to a control group (Rim et al. 2008). In a cluster-randomized, community-based, 2-month effectiveness trial in the Kyrgyz Republic, the intervention group showed a significant increase in Hb concentrations and a significant decrease in the prevalence of anaemia when compared to the control group. The additional daily iron dose over 2 months was 12.5 mg as microencapsulated FeFum (Lundeen et al. 2010). Relatively low and infrequent use of Sprinkles containing 12.5 mg iron (0.9 sachet/week) was still associated with decreased rates of anaemia and ID in a 12-month follow-up when compared to a control group (Suchdev et al. 2012). In another study, haematological response (prevalence of anaemia) to a flexible administration of 60 Sprinkles sachets (12.5 mg iron as FeFum) over 4 months was better than a daily administration over 2 months. This was explained by the better adherence to MNP administration and better acceptability when flexible administration was allowed (Ip et al. 2009). However, for moderately to severely anaemic children, daily MNPs containing 10 mg iron were more effective in improving Hb concentrations and reducing anaemia prevalence than only two sachets of MNPs per week (Kounnavong et al. 2011). Compared with drops of $FeSO_4$ (10 mg iron), daily Sprinkles (12.5 mg iron as FeFum) showed no difference in changes of anaemia prevalence, but had significantly less benefit on Hb and SF concentrations than the drops. The authors argued that absorption from $FeSO_4$ is generally higher than from FeFum and that the $FeSO_4$-drops were administered directly without mixing it with food (Samadpour et al. 2011). Some studies tried to use lower concentrations of iron (2.5 mg) when using NaFeEDTA. While two of these studies found only a small improvement in iron status (Ndemwa et al. 2011; Troesch et al. 2011), a RCT in Kenyan preschool children was more efficient reducing the prevalence of anaemia, ID and IDA and improving Hb, SF and TfR concentrations (Macharia-Mutie et al. 2012). It has to be stressed that the study conducted by Troesch et al. used NaFeEDTA in combination with exogenous phytase.

MNPs have several advantages. As they can be mixed with home-made foods, they do not require any change in food practices and they can provide the daily dose of iron and further micronutrients regardless of the quantity of complementary food eaten by the child. Sachets of MNPs are easy to use and convenient even for illiterate people. Transport, storage and distribution are facilitated by the lightweight sachets and the long shelf-life of the MNPs even in hot humid conditions. In addition, the costs of MNPs are note excessive (~0.035 US$/sachet/daily ration) (Sprinkles Global Health Initiative 2012). The main cost is the sachet packaging. MNPs appeared to be cost beneficial when lifetime earnings through MNPs feeding were considered (Sharieff et al. 2008). MNPs with low iron doses can be used in malaria endemic areas without safety concerns (Troesch et al. 2011). The main disadvantage of MNPs is that they do not provide the macronutrients which are also often insufficient in indigenous complementary foods for young children in developing countries (Gibson et al. 1998). Furthermore, the delivery of MNPs and maintaining adherence to MNPs is challenging. In a refugee camp in Kenya, the uptake of MNPs dropped from almost 100% to a low of 30% and remained at a level of 45–52% despite increased social marketing efforts. The reasons were inadequate social marketing prior to the program and scarce training of community health workers. This resulted in insufficient communication about the health benefits and the use of MNPs to the beneficiaries. Furthermore, issues with the packaging, such as the lack of information about halal or haram and the hurting of religious feelings by the cartoon logo, decreased reliance to the MNPs. Programs using MNPs, therefore, need a careful design which ensures the involvement and commitment of all stakeholders, always considering cultural background (Kodish et al. 2011).

3.1.2.3 Fortified lipid-based spreads

Fortified lipid-based spreads, also called LNSs, are high-viscosity-fat products produced by mixing dried powdered ingredients including a micronutrient premix with a vegetable fat chosen for its viscosity (Briend and Solomons 2003; Dewey et al. 2009b). They were initially developed for severely malnourished children, but may also offer an effective and convenient food-based way for the prevention of micronutrient deficiencies and the promotion of growth (Phuka et al. 2011). LNSs can be eaten alone as a snack or preferably mixed with traditional porridges used for complementary feeding (Briend 2001).

The main advantage of LNSs is that they provide energy, fat (including essential fatty acids) and protein in addition to micronutrients, and therefore may have a beneficial effect on growth. In developing countries, traditional complementary feeding is often based on bulky porridges made of a cereal alone or mixed with legumes. These bulky foods are not only low in micronutrients but also in energy. By mixing LNSs with a traditional porridge, energy and micronutrient density of the porridge is improved. In a 6-month intervention trial using LNSs, MNPs, and iron tablets, only the LNSs had a positive effect on weight and length gain in children. Compared to the control group, all three interventions had a beneficial effect on the percentage of infants, who were able to walk independently at 12 months of age, with the greatest impact seen in the LNS group (Adu-Afarwuah et al. 2007). Positive effects on weight (Lin et al. 2008) or length (Phuka et al. 2009) gain were also observed in other studies using LNSs.

Regarding improvement of iron status, data on LNSs used for *in-home* fortification is limited to two studies. In the first study, fortified spreads based on milk powder or defatted soy flour mixed with peanut butter, sugar and oil tended to increase Hb concentrations in moderately underweight Malawian children ($P = 0.06$) (Kuusipalo et al. 2006). This study had eight intervention groups with the following food fortification scheme over a period of 12 months: nothing, 5, 25, 50, or 75 g milk-based spread per day (11 mg iron in all portions) or 25, 50, or 75 g soy-based spread per day (12–13 mg iron in all portions). Information about the type of iron fortificant is not available. The second study compared the use of LNSs (9 mg iron as $FeSO_4$) with MNPs (12.5 mg iron as FeFum), iron tablets (9 mg iron as $FeSO_4$) and a randomly selected control group of infants assessed at study endpoint. After 6 months of daily intervention, all 3 intervention groups had significantly higher SF and significantly lower TfR concentrations than the control group. Compared to the control group, Hb concentrations were significantly higher in the LNS and iron tablet group but not in the MNP group. Prevalence of IDA was significantly lower in all intervention groups when compared to the control group (Adu-Afarwuah et al. 2008).

LNSs usually have low water content (<2%), thus they can be easily stored in tropical environments without risk of pathogenic bacterial proliferation (Briend 2001). However, storage is limited due to oxidation causing a rancid taste and leading to decreased vitamin concentrations. A prolonged shelf-life can be obtained by packaging under nitrogen into aluminium foil wrappings, but this type of packaging is often not available when producing LNSs locally (Briend and Solomons 2003). Current LNSs are produced in industrialized countries and are typically made from vegetable oil, peanut paste, milk powder, sugar, and a micronutrient premix (Phuka et al. 2011). Acceptability of such LNSs was good in Ghana (Adu-Afarwuah et al. 2008) and Niger (Tripp et al. 2011). However, the use of milk powder and rapeseed (canola) oil contributes to the reasonably high costs of LNSs at present time (0.09–0.30 US$/daily ration). Highly concentrated LNSs in which imported milk powder and canola oil is replaced by domestic products, such as defatted soybean flour or soybean oil, would be cheaper and can be locally manufactured.

Soybean oil is readily available in many African countries and essential fatty acid composition is only slightly inferior to canola oil (Phuka et al. 2011). Replacing milk powder with an ingredient such as defatted soybean flour that has poorer protein quality, is a critical issue as there is some evidence that milk peptides promote linear growth in children (Hoppe et al. 2006; Hoppe et al. 2008). Phuka and colleagues investigated the acceptability of four LNSs including low-cost types (0.04–0.08 US$/daily ration) without milk powder and with a different oil source. All LNSs were well accepted without differences between the different types. The use of soy protein raises a greater concern regarding allergenicity and decreases mineral bioavailability due PA. Potent antigens can also be found in peanut paste and additionally toxicity of aflatoxin contamination is a serious issue when using peanut paste (Briend and Solomons 2003).

3.2 Biofortification

Biofortification refers to the development of micronutrient-enhanced staple crops by conventional plant breeding practices or by modern biotechnology (Nestel et al. 2006). It can be sustainable and cost-effective approach to combat micronutrient deficiencies (Qaim et al. 2007). Both conventional plant breeding and genetic engineering involve altering the genotype of the targeted crop with the objective of developing plants carrying genes that support the accumulation of bioavailable micronutrients, but the means of reaching this purpose are different in the two approaches.

HarvestPlus, the global leader in the development of biofortified staple crops, runs several biofortification programs employing more than 200 agricultural and nutrition scientists around the world (Nestel et al.

2006). They are currently evaluating the potential of a selection of staple crops (pearl millet, rice, wheat, cassava, maize, sweet potato, beans) for provitamin A, iron or zinc biofortification programs in Africa, Asia, and Latin America (Bouis et al. 2011). The following three chapters are mainly focused on iron biofortification of staple crops. They discuss the implementation, the advantages and the limitations of iron biofortification.

3.2.1 Implementation of iron biofortification

Successful implementation of biofortification requires several activities. HarvestPlus developed a pathway of impact flow-chart for these activities (Figure 10) including three phases: 1) discovery, 2) development, and 3) delivery (HarvestPlus 2011a).

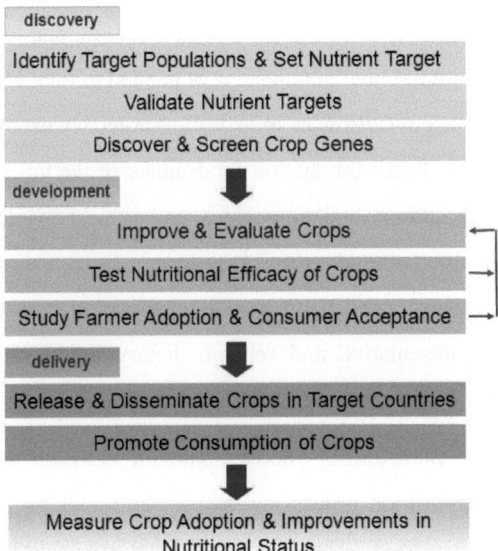

Figure 10: Steps of the HarvestPlus impact pathway for biofortified crops (HarvestPlus 2011a).

3.2.1.1 Discovery phase

The discovery phase starts with the identification of the targeted populations and with setting nutritional breeding targets. The decision where iron-biofortified varieties should be targeted is based on cropping and food consumption patterns, the incidence of ID with or without anemia, and preliminary benefit-cost analysis (HarvestPlus 2011a). The targeted populations are not necessarily restricted to only one country and spillover effects to other countries have to be taken into account (Ortiz-Monasterio et al. 2007). Information about scientific capability and institutional strength of local agricultural research and extension system, about political stability, and about the strength of supporting governmental and non-governmental enabling institutions were gathered during this first step (Bouis et al. 2011).

Identification of target population is often complicated due to fragmentary or missing data from national health surveys needed to identify populations affected by ID (Zapata-Caldas et al. 2009). Target levels of iron in the biofortified crop are based on data on food intake in the target population, iron losses during storage and processing, and iron bioavailability in the foods made of the targeted crop. Assessment of consumption of the targeted crop in the targeted population is challenging due to the limited availability of representative and reliable dietary intake data (Hotz and McClafferty 2007).

In a second step, nutritionists validate data on nutrition and ID. They investigate the staple food processing, storage and cooking practices, and the inclusion of other foods. These are factors that can lead to different iron concentrations and bioavailability in the food prepared from the targeted crop (Bouis et al. 2011). Particularly, processing steps, such as milling or polishing, can lead to a significant decrease in iron concentrations in the targeted crop (Pfeiffer and McClafferty 2007). The data collected by the nutritionists guide the plant breeders in confirming or upgrading their breeding objectives (HarvestPlus 2011a).

Still during the discovery phase, plant breeders identify whether the genetic variation of the targeted crop is sufficient to breed for the iron level needed in the target population (Bouis et al. 2011). HarvestPlus has access to the huge reservoir of staple crops germplasm provided by the global germplasm banks of the Consultative Group on International Agricultural Research (CGIAR). If accepted, upstream transgenic research is also considered to reach iron targets that are too difficult to obtain through conventional breeding (HarvestPlus 2011a).

3.2.1.2 Development phase

Crop improvement and evaluation are the first steps during development phase. The promising lines identified by the plant breeders are now mapped by their genotypic differences. New varieties are developed by crossing promising lines and selecting those which perform well in farmer's field and meet farmer's expectations while still providing better iron supply to the targeted population (Grusak and Cakmak 2005). To assess genotype and environment (G x E) interactions, the performance of the recently developed iron-biofortified varieties is tested across different crop growing scenarios. There is evidence that the variability of iron in the germplasm depends on the genotype, the environment and G x E interactions but the impact of the various factors varies among crops (Oury et al. 2006).

Genetic variability for varieties with high iron concentrations is limited for some crops. Additionally, iron bioavailability from foods prepared from the iron-biofortified crop is often not optimal. Thus, plant breeders do not only aim for crops with high iron concentration but also for crops with low/high concentrations of inhibitors/enhancers. Such combined approaches make success of iron biofortification more likely (Nestel et al. 2006). PA in foods made from iron-biofortified crops can be decreased by transgenic plants overexpressing phytase. Some studies successfully introduced and expressed an *Aspergillus (A.)* phytase gene in rice (Lucca et al. 2001), maize (Drakakaki et al. 2005) and wheat (Brinch-Pedersen et al. 2006). Due to the low heat stability, the phytase was degraded during food preparation and did not reduce PA significantly (Lucca et al. 2001). Nevertheless, such a cereal overexpressing phytase can be of interest for preparations where the flour is mixed with water before the heat treatment (e.g. traditional fermentations). Lower PA concentrations were also reported in low-PA mutants (Lpas) of wheat (Guttieri et al. 2003), rice (Larson et al. 2000), maize (Raboy et al. 2000), barley (Larson et al. 1998), sorghum (Kruger et al. 2012) and beans (Campion et al. 2009). *In vitro* iron availability from fermented porridges made of non-tannin Lpa sorghum (213 mg PA/100 g) was increased compared with fermented porridges made of non-tannin normal PA sorghum (395 mg PA/100 g). However, *in vitro* iron availability did not differ when the same sorghums were compared in unfermented porridges (794 mg vs. 1165 mg PA/100 g). This suggests that a substantial decrease in PA is needed to improve iron bioavailability in certain crops (Kruger et al. 2012). In a human study, fractional iron absorption from Lpa maize was 50% higher than from non-genetically modified maize (Mendoza et al. 1998). However, iron bioavailability from Lpa maize fortified with iron (1 mg/serving) was not different than that from iron-fortified normal PA maize (Mendoza et al. 2001). At the moment, the two major drawbacks of many Lpa crops are reduced yield and decreased seed germination (Raboy 2001; Guttieri et al. 2006).

Crops with increased concentrations of AA or cysteine-rich endogenous protein might have the potential to improve iron bioavailability from various plant foods. Increased AA concentration has been reported in genetically modified maize (Chen et al. 2003). Further research is needed to evaluate whether increased AA concentrations in other cereals are feasible and whether it would have a positive impact on iron bioavailability. AA is sensitive to heat and losses during cooking have to be considered when setting the target levels. Researchers have been succeeded to overexpress a cysteine-rich endogenous protein in rice and increase its concentration 7-fold (Lucca et al. 2001). It has to be studied if such an increase is sufficient to improve iron bioavailability from rice foods.

Plant breeding also includes the application of fertilizers which provide essential nutrients to the plant through nutrient-enriched soil. However, this approach is only applicable for a few nutrients, such as zinc, iodine or selenium, but ineffective for iron. Most of the iron is rapidly oxidized in the soil and iron has low mobility in the phloem (Graham et al. 2001). Furthermore, the total mineral concentration of iron in the most soils is expected to be sufficient to breed iron-dense crops (White and Broadley 2009).

Testing the efficacy of the iron-biofortified crop is one of the key steps in the development phase. If efficacy cannot be shown, it is unavoidable to go back to the steps before and reconvene the improvement of the biofortified crop. Before testing efficacy, studies investigating the bioavailability of the iron bred into the crop *in vitro* (e.g. Caco-2-cell studies) or preferably *in vivo* (stable or radioactive isotope studies) can give important information (Kennedy et al. 2003; Hotz and McClafferty 2007) (see also chapter 3.5: Evaluating iron bioavailability from (bio)fortified foods). These studies can examine the assumption made by nutritionists and plant breeders under the discovery phase. If promising results were obtained from these preliminary measures, the newly developed biofortified variety is eventually investigated in an efficacy trial in human subjects. The efficacy trial is able to show whether an intervention (in this case the biofortified crop) results in the expected outcomes under idealized conditions (Hallfors et al. 2006).

In the last step of the development prior to the release of the iron-biofortified crop, researchers have to study the criteria that affect whether farmers and consumers will adopt and consume the biofortified variety or the products made of it. If the biofortified variety fails to fulfil one of these criteria (e.g. same or accepted organoleptic properties compared with the non-biofortified variety), researchers have to revert to the steps where they evaluate and improve the biofortified variety (HarvestPlus 2009b). Effectiveness trials are useful tools to approach such issues. Compared to efficacy trials, these trials are testing the effects of the crop under conditions that simulate reality more precisely (Gartlehner et al. 2006). The studied population is more broadly defined and the biofortified crop is prepared and eaten in traditional ways within the usual household environment. Standardization only occurs in terms of access and availability of the biofortified crop among the target population.

3.2.1.3 Delivery phase

The delivery phase starts with the registration of the biofortified crop according to the regulations of the targeted countries. Usually, registration requires evidence that the variety is new, distinguishable and adding value (Bouis et al. 2011). The next step will be the development of strategies on how to multiply biofortified planting material and the creation of a market for the biofortified seeds and crops (Bouis and Welch 2010). This includes dissemination and promotion of the biofortified crop among farmers but also achieving behavioural changes at policy, institutional, community and individual levels (HarvestPlus 2009b). In the past, delivery of newly developed varieties has failed due to insufficient seed multiplication and distribution (David et al. 2002; Chirwa et al. 2006). The successful delivery of biofortified beans included: 1) informal seed multiplication using smallholder farmers, 2) informal seed distribution by grocery shops, rural traders, extension agents, health clinics, and non-governmental organizations (NGO), 3) intense publicity through promotional materials like posters, leaflets, brochures and radio messages, and 4) informal outlets such as farmers, NGO, extension agencies, village traders and various other institutions (Chirwa et al. 2006). Iron biofortification is successfully implemented when the iron status of the targeted population is improved in baseline and follow-up surveys after delivery. Such surveys can be similar to effectiveness trials and include investigations on the number of farming households that have adopted the biofortified crop (HarvestPlus 2011a).

3.2.2 Advantages of iron biofortification

Compared to other approaches combating ID, biofortification can have multiple advantages. First, it is a food-based approach which, in contrast to dietary diversification, does not require behavioural changes from the target population. Staple crops are predominant in the diet of the poor, and therefore iron biofortification will implicitly target all family members of low-income households which are at highest risk for ID (Tanumihardjo et al. 2008). Second, once the iron-biofortified crop has been developed, the recurrent expenditures are low and the germplasm can be shared internationally. These multiplying aspects of plant breeding across time and distance make biofortification more cost-effective than traditional fortification for which on-going buying of iron fortificants and food processing can be very cost-intensive (Qaim et al. 2007). Third, after successful implementation, the iron-biofortified crop will continue to be grown and consumed year after year, even if government attention and international funding for ID issues is weak (Nestel et al. 2006). Fourth, undernourished populations in relatively remote rural areas have limited access to commercially marketed fortified foods but can be reached relatively easily by biofortified crops. This makes iron biofortification and conventional iron fortification highly complementary (Bouis and Welch 2010).

3.2.3 Limitations of iron biofortification

3.2.3.1 General limitations of biofortification

To completely exploit the above mentioned advantages, certain factors have to be considered during development of biofortified crops (Welch and Graham 2004). The potential benefits of iron biofortification are not equivalent among different age groups and among men and women. It depends on the amount of staple crop consumed, the prevalence of existing ID, and the iron requirements as affected by daily losses of iron from the body and special periods with higher iron needs such as growth, pregnancy, and lactation (Hotz and McClafferty 2007). Biofortification alone is most likely not sufficient to meet iron requirements of young children under 2 years of age, who have particularly high iron needs but relatively low intakes of staple foods (Bouis et al. 2011). It is therefore important to carefully identify the targeted crop and the target population.

Iron biofortification might improve iron status of those individuals who are less iron deficient and possibly maintain iron status of all the population at low cost, but is not an appropriate approach to combat ID with severe anaemia in severe acute malnutrition. In addition, an appropriate target level of iron in the biofortified crop has to be set, and bioavailability of the additional iron has to be investigated (Ortiz-Monasterio et al. 2007). It is also important that the productivity and yield of the iron-biofortified crop is maintained or augmented to assure farmer acceptance. Furthermore, the appearance of the iron-biofortified crop and its sensory properties after processing are crucial for consumer acceptance (Nestel et al. 2006).

3.2.3.2 Acceptance of transgenic plants

Due to the weak acceptance of transgenic plants, biofortification research programs are mainly focused on the development of iron-biofortified crops by conventional plant breeding. Modifications of genes creating genetically modified organisms (GMOs) has been successfully applied to increase iron concentrations in crops which do not have the genetic potential to meet desirable iron levels by conventional plant breeding (Borg et al. 2009). The main advantages of GMOs are their rapid development and the stable expression due to the precise introduction of single genes in the targeted plant. Therefore, less breeding generations are needed to obtain the biofortified crop by genetic engineering compared to conventional plant breeding. On the downside, the development of GMOs is often linked with patented or patentable inventions making them poorly accessible for researchers in developing countries and unaffordable for farmers (Pardey et al. 2001). GMOs face various regulatory and political restrictions and also social or ethical considerations (WHO 2005) which have resulted in a strong opposition to them by some groups. Health concerns, especially the initial and controversial use of antibiotic resistance markers, have deepened the opposition to GMOs (Seralini et al. 2007). This is because it is theoretically possible that transgenes survive human digestion and reach the colon where they can transfer to the intestinal microbiota and so installing antibiotic resistance genes (Netherwood et al. 2004). However, it has now been shown that the transfer of genes occurs only at undetectable frequencies and the practice of using antibiotic resistance genes has been started to phase out (Demaneche et al. 2008). GMOs are also suspected of exhibiting certain toxicity, allergenicity, and carcinogenicity or to affect human health indirectly through their negative impact on the natural environment e.g. invasion of natural habits, decrease of biodiversity, horizontal gene transfer (Conner et al. 2003), and through economic, social or ethical factors (WHO 2005).

3.3 Enhancing bioavailability of (bio)fortification iron

Several approaches are known to enhance the bioavailability of fortificant iron or native iron bred into crops through biofortification: 1) adding AA or other organic acids, 2) the addition of disodium EDTA (Na_2EDTA), 3) reduction of PA, and 4) reduction of PPs. Regarding fortificant iron, there is a fifth approach, namely the changes of physical properties of iron compound (e.g. particle size). The removal of PPs, the physical properties of iron compounds and the effect of organic acids have already been described in previous chapters (2.1 Iron absorption enhancers, 2.2.2.4 Improved iron bioavailability through reduced polyphenol concentration, 3.1.1.5 Encapsulated, micronized and nanostructured iron compounds) and are, therefore, not further discussed in the following chapters

3.3.1 Adding ascorbic acid

Adding AA has been shown to substantially increase the bioavailability of most iron compounds and has been practiced widely in the food industry (Hurrell 2002a). However, there is some uncertainty about the influence of AA on the absorption of certain iron compounds such as FeFum or NaFeEDTA. Two human studies reported no or only moderate improvement of FeFum absorption, when adding high amounts of AA (Hurrell et al. 1991; Davidsson et al. 2000). A recently published study reported no beneficial effect on iron absorption when AA was added to a NaFeEDTA-fortified maize meal compared with the same meal without AA (Troesch et al. 2009). Adding AA is a suitable approach for many packaged processed foods, but is not recommended for staples and condiments because of stability issues. Losses during storage can be unacceptably excessive under hot and humid conditions, and although sophisticated packages can mainly prevent degradation, they are most likely too expensive for many applications, especially in developing countries.

For most foods, it is recommended to add AA in an AA:Fe of 2:1 (= 6:1 weight ratio). Higher AA:Fe (4:1) is appropriate for foods containing high concentrations of PA. Such molar ratios are necessary to achieve a 2- to 3-fold increase of iron absorption in adults and children (Hurrell 2002a). Discrepancies are found regarding the long-term effect of AA on iron status. In a study with preschool children consuming plant-based diets, iron status (Hb, SF) significantly improved after a 8-week supplemental period with AA (Mao and Yao 1992). In another study with iron-deplete women, AA supplementation only had a beneficial effect on Hb concentrations (Hunt et al. 1990). Additionally, a prospective cohort-study in non-pregnant Mexican women showed that AA intake was positively correlated with SF and Hb concentrations (Backstrand et al. 2002). In the Framingham Heart Study, a longitudinal study on heart disease risk factors in elderly, dietary AA was positively associated with SF concentrations (Fleming et al. 1998). In a recent 4-months intervention trial, the addition of kiwifruit rich in AA, lutein and zeaxanthin to a breakfast cereal fortified with $FeSO_4$ (16 mg/serving) improved iron status (SF, TfR) in women with low iron stores compared to a control group consuming the breakfast cereal with banana (Beck et al. 2011). On the other hand, several studies have reported no effect of AA supplementation on iron status. In a study conducted with iron deficient women, AA supplementation did not significantly improve SF concentrations ($P = 0.06$) (Hunt et al. 1994). The authors argued that iron absorption was maybe adapted to the presence of inhibitors and enhancers, independent of iron status (Hunt 2003). No improvement in SF concentrations was also reported in two other studies with less iron-deplete or even iron-replete subjects (Cook et al. 1984; Malone et al. 1986). Due to the differences among sample size, duration of intervention, baseline iron status, AA:Fe, age of subjects and type of diet, the interpretation and comparison of all these studies is challenging. Further research is needed to assess the efficacy of AA as a means to improve iron status, particularly of iron-depleted individuals (Teucher et al. 2004).

3.3.2 The use of disodium EDTA

A valuable alternative to AA is Na_2EDTA which is a permitted food additive in many countries. Unlike AA, it is not an essential nutrient and it is stable during processing and storage. In the stomach at low pH, Na_2EDTA is able to chelate iron, and thus protect it from binding to inhibitors such as PA and PPs (WHO 2006c). It can improve absorption of both native food iron and soluble fortificant iron, but not that of relatively insoluble compounds such as FeFum (Davidsson et al. 2002), FePP (Hurrell et al. 2000b) or elemental iron (Fairweather-Tait et al. 2001). Numerous studies have demonstrated the positive effect of Na_2EDTA on iron absorption from $FeSO_4$ (Viteri et al. 1978; Macphail et al. 1994; Hurrell et al. 2000b; Davidsson et al. 2001b). For a 2- to 3-fold increase in iron absorption, recommendations suggest a Na_2EDTA:iron molar ratio between 0.5–1.0 (= 3.3:1-6.6:1 weight ratio) in the food with added Na_2EDTA (WHO 2006c) but results are sometimes inconsistent. Some studies stated higher iron absorption from $FeSO_4$ at molar ratios <1.0 (Macphail et al. 1994; Davidsson et al. 2001b), while another study reported that a molar ratio <1.0 was only effective in test meals with low PA concentrations. In meals with high PA concentrations, a molar ratio of 1.0 was the most effective (Hurrell et al. 2000b). In contrast to the results presented above, an early radioisotope study in men showed a strong inhibitory effect of EDTA on iron absorption from an American composite meal at molar ratios >0.5:1 (Cook and Monsen 1976b).

3.3.3 Reduction of dietary phytic acid

Reduction of PA is a useful approach to substantially increase absorption from iron-(bio)fortified foods, particularly from cereal-, legume- and soy-based complementary foods or infant formulas. PA inhibits iron absorption even at low concentrations. Therefore, if applicable, complete reduction is suggested to assure a meaningful increase in iron absorption (see chapter 2.2.1.1: Interaction of phytic acid with iron). Dietary PA can be decreased by breeding Lpa crops (see chapter 3.2.1.2: Development phase). Other methods to reduce PA by removal include milling, water washing of milled flour, and dialysis and ultra-filtration of acid/salt or alkali treated legume protein isolates. Further approaches to reduce PA are soaking, malting, germination and fermentation of cereals and legumes (Hurrell 2004b) (see also chapter 2.2.1.2: Processing steps affecting phytic acid concentration in sorghum and millet foods). During such processing steps, which are often important in the preparation of traditional foods in developing countries, PA is degraded by activation of native phytases which remove the iron-binding phosphate groups. Native phytase activity is highest in whole wheat and whole rye but generally low in other cereals and in legumes. In complementary foods, PA can be degraded during manufacture by formulating the composition of the food to contain cereals, such as wheat, with a high native phytase activity (Egli et al. 2003). In addition to the activation of native (endogenous) phytases, fermentation also provides phytases from yeasts, moulds and lactic acid bacteria (Sharma and Kapoor 1996; Songre-Ouattara et al. 2008).

A more novel approach is adding exogenous phytase (purified from bacteria, fungi, or plants) during processing. Iron absorption from soy-formula with no PA, due to the use of phytase from *A. niger*, was doubled in infants when compared with untreated formulas (Davidsson et al. 1994). Successful dephytinization needs a processing step during which the cereal mixture and the phytase are held in an aqueous medium for an appropriate time at pH and temperature conditions favourable for the phytase. In developing countries, the low-cost extruded cereals for complementary feeding do not include such an aqueous phase, thus such dephytinization is still a challenge (Hurrell 2004b).

A slightly different approach, already widely used in animal feeding, is the addition of phytase active at gastric pH. This approach has gained much interest in recent years and is considered as a promising approach to increase iron bioavailability from complementary foods as well as other foods. The phytase has temperature and pH optima adapted to the intestinal tract and is thereby able to degrade the PA during digestion (Troesch et al. 2009). Phytase that is active at gut pH has been shown to increase iron absorption in two studies when added to a meal just before consumption (Sandberg et al. 1996; Troesch et al. 2009). In the more recent study by Troesch et al. 2009, the addition of phytase to a whole-maize porridge significantly improved iron absorption from NaFeEDTA and $FeSO_4$, with or without added AA. The same phytase has been included in a low-iron MNP used in a short-term efficacy trial in school children (Troesch et al. 2011).

The use of phytase in fortification programs may increase efficacy of low iron concentrations in highly inhibitory complementary foods. The current phytases are usually obtained from a genetically modified *A. niger*, but the phytase itself is purified and does not contain any genetically modified material. While legal issues, such as the restricted use of GMO derived material by national legislation in some countries, still remain, the usage of the phytase produced from genetically modified *A. niger* has recently been approved by JECFA. JECFA allocated an ADI "not specified" for the use of *A. niger* phytase in applications such as LNSs, FBFs, MNPs, fortified flours, breakfast cereals and beverages (WHO 2012). Patents, such as for DSM phytase Tolerase™ 20000G, might limit their use in locally produced foods in developing countries. Costs can be an issue, especially if sophisticated packaging is needed to assure storage stability in hot and humid conditions. More research is needed to gather more information on potential ways to increase the stability, the activity and the efficacy of the phytase active at gastric pH and added to different food matrixes.

3.4 Iron (bio)fortification of sorghum and millets – state of the art

Conventional iron fortification of food vehicles, including wheat and maize flours, cereal-based complementary foods, fish and soy sauce, and milk, is already established in many parts of the world (WHO 2006c). Mass iron fortification of cereal flours is currently restricted to wheat and maize, for which evidence based recommendations have been established (WHO et al. 2009). Sorghum and millets have been neglected as potential vehicles for iron fortification despite being important staple cereals for many people. This is linked to the general neglect of sorghum and millets in research and to the lack of a sorghum and millet industry in the respective countries. However, there are some human iron absorption studies using sorghum-based meals (Gillooly et al. 1984; Hurrell et al. 2003). These studies reported that high concentrations of PPs in some sorghum varieties strongly inhibit iron absorption, suggesting that not only PA but also PP concentrations in sorghum and some millet varieties could be an issue in potential iron fortification programs. This is in contrast to iron fortification of wheat and maize which do not contain significant amounts of PPs.

Iron fortification of sorghum and millets has been investigated in a few *in vitro* dialyzability trials using the method described by Luten et al. 1996. Low-tannin white sorghum and high-tannin red finger millet were fortified with 6 mg iron as FeFum per 100 g flour along with Na_2EDTA. Dialyzable iron was measured directly in the flours and as well in food products made of the fortified flours. Compared with unfortified flours and their products, fortification significantly increased the total amount of dialyzable iron. There was no difference in iron dialyzability between low-tannin sorghum and high-tannin finger millet indicating that the tannins did not influence dialyzable iron (Tripathi et al. 2012). In the same study, the shelf-life of the fortified flours was adequate up to a period of 60 days and fortification did not adversely alter the sensory quality of the food products. In another study, the same research team investigated FePP as an iron compound in finger millet flours and compared it to FeFum (fortification levels = 6 mg iron/100 g flour). Both iron compounds were found to provide equal amounts of dialyzable iron. Adding Na_2EDTA at levels equimolar to the added iron, significantly increased the dialyzability of iron from the fortified flours (Tripathi and Platel 2011). Similar findings were found in a study investigating iron accessability from FeFum-fortified finger millet after different processing steps (Oghbaei and Prakash 2012). *In vitro* dialyzability trials can provide important insights in terms of iron fortification but careful evaluation of results is needed. It is for example possible that PPs form small soluble complexes with iron which can pass a dialysis membrane but are not absorbable.

At the moment, several crops are undergoing an iron biofortification process in various HarvestPlus programs. Iron biofortification of beans in Rwanda is the most advanced program with some prelaunch seed multiplication (Bouis et al. 2011). Efficacy of iron-biofortified crops has been tested with rice in a 9-month feeding trial with Filipino women. No significant increase in body iron, SF and Hb concentrations was detected compared to the control group. Differences were only reported in a subgroup of non-anaemic women which had increased body iron and SF concentrations compared to the control group of non-anaemic women. It has to be stressed that, due to losses during processing, the additional iron intake from the iron-biofortified rice was <2 mg per day (Haas et al. 2005) whereas iron fortification programs target 6 mg additional iron per day (Hurrell et al. 2010).

Compared with conventional fortification, the work on biofortification of sorghum and millets has rapidly moved forward over the last years. This is not surprising as one of the closest HarvestPlus stakeholder, the International Crops Research Institute for the Semi-Arid Tropics (ICRISAT), has it's headquarter in India where the two grains play an important role in the daily diet. Biofortification of pearl millet has advanced farthest, and strategies for the release are already in the pipeline

(HarvestPlus 2011b). The aim of HarvestPlus is that 10 years after release, 28 million people in India will be consuming new iron-rich pearl millet (HarvestPlus 2009a). HarvestPlus has identified India as the first target and Mali and Niger as additional recipient countries. The initial breeding target is set at 7.7 mg iron/100 g pearl millet to provide 30% of mean daily iron requirement through normal consumption habits. This is based on the following assumptions using women as a reference: 300 g pearl millet intake per day, 90% retention after processing and cooking, and 5% bioavailability (HarvestPlus 2009a; HarvestPlus 2009c). Compared with the average iron concentration measured in pearl millet (4.7 mg iron/100 g pearl millet), the iron concentration in the biofortified variety is increased by almost 65%. The potential of genetic enhancement for increased iron levels in pearl millet is high (Velu et al. 2007; Velu et al. 2008; Rai et al. 2012), and therefore the targeted iron level has already been achieved. However, iron concentration was negatively associated with grain yield in some trials. It is, therefore, necessary to select varieties that are not compromising grain yield (Rai et al. 2012). Iron and zinc concentrations are positively correlated in pearl millet. This opens the possibility to develop pearl millet varieties high in iron and zinc to combat iron and zinc deficiency (Gupta et al. 2009; Rai et al. 2012). Performance testing of the new pearl millet varieties in the field is on-going as well as measurements of nutrient retention during processing. Furthermore, bioavailability and efficacy studies to evaluate impact on iron status are on-going or planned (HarvestPlus 2011b). A baseline study in the targeted communities is also planned to measure impact on nutritional status after the delivery of high iron pearl millet into the diet (HarvestPlus 2009a).

In a recent study, iron bioavailability from two biofortified pearl millet varieties, developed by ICRISAT Mali through conventional breeding, was estimated using the PA:Fe as an indicator. The authors reported that the iron bioavailability in the biofortified varieties is expected to be higher than from the control variety, even after decortication. However, large losses of total iron (>40%) were reported after decortication (Hama et al. 2012). Screening for the genetic diversity of finger and foxtail millet showed some moderately good prospects for high iron concentration breeding. The variability in iron concentration exists but is inferior to pearl millet (Upadhyaya et al. 2011a; Upadhyaya et al. 2011b).

Regarding sorghum, the prospects for an iron-biofortified variety through conventional breeding are more limited. In a study screening 84 sorghum lines, the iron concentration ranged only between 2.0 and 3.7 mg iron/100 g grains (Reddy et al. 2005). In contrast, a more recent study showed better variability of iron in sorghum ranging from 2.1 to 6.1 mg iron/100 g grains (Ashok Kumar et al. 2009). Preliminary data suggest that sorghum has good PA variability ranging from 380 to 1350 mg PA/100 g sorghum (Reddy et al. 2005). Therefore, if further research confirms the better genetic variability in iron concentration, breeding for high iron and low PA sorghum might deliver sorghum varieties with high iron bioavailability. The Africa Biofortified Sorghum (ABS) project is developing a transgenic sorghum with increased levels of essential amino acids, especially lysine (80–100% increase), improved protein digestibility, and increased vitamin A, vitamin E, iron (50% increase), and zinc (35% increase). The primary purpose of this sorghum is better protein nutrition (Taylor et al. 2012) and its application is still restricted in many countries by regulations against GMOs.

3.5 Evaluating iron bioavailability from (bio)fortified foods

Successful iron (bio)fortification requires sufficient iron bioavailability from the (bio)fortified vehicle. Therefore, prior to human efficacy or effectiveness trials, methods to predict iron bioavailability are needed to estimate the potential impact of the (bio)fortified vehicle. These methods include absorption studies using isotopes, animal models, and *in-vitro* methods such as solubility tests, dialyzability tests and cell models. Another earlier method is the chemical balance technique which measures the difference between the amount of a mineral eaten and the amount in the faeces. This method, however, is not sensitive, not precise and time-consuming, and it cannot be used to gather information about iron absorption from different meals (Hallberg 1981; Consaul and Lee 1983). Similar to the chemical balance technique, algorithms for calculating absorption and bioavailability of dietary iron can be used for predictions. However, algorithms require detailed information about meal composition and its variation over a long period of time, and they can only be applied to whole diets (Hallberg and Hulthen 2000).

3.5.1 *In-vitro* methods and animal models

Iron compounds for fortification can be evaluated *in-vitro* using solubility tests. In these tests, a known amount of iron is given into a solution of diluted acid with fixed pH, usually similar to gastric juice pH. After a certain time, solubility is measured relatively to the total amount of added iron. These tests are rapid and inexpensive, but with regard to native food iron, their usefulness to predict iron bioavailability is limited (Miller et al. 1981; Fairweather-Tait et al. 2007). Dialyzability is a further developed *in vitro* method involving a two-stage simulated digestion process (pepsin and pancreatic digestion) and the use of a dialysis membrane with a selected molecular weight cut-off. It has been reported that compared to human absorption studies, dialyzability tests usually, but not always, predict the correct direction of response. However, the measured magnitude of the effect is rarely similar to human studies (Fairweather-Tait et al. 2005). Further limitations are the low reproducibility (Luten et al. 1996) and the fact that the membrane selects the passing iron-complexes only by size and not by bioavailability (Fairweather-Tait et al. 2005).

The most sophisticated *in vitro* method to test iron bioavailability is the Caco-2 cells approach. The analysis of iron uptake in these human epithelial cell lines is commonly preceded by a two-step digestion to simulate conditions in the gastrointestinal tract. In addition, the Caco-2 cells allow the analysis of iron transport by culturing the cells on inserts. With some exceptions, the results from a combined approach using digestion and Caco-2 cells often correlate with human iron absorption studies in prediction of direction of response, but the magnitude differs from human studies (Sandberg 2010). Low reproducibility has been reported mainly due to the fact that Caco-2 methodologies vary significantly between different laboratories (digestion steps, cell culture quality, etc.) (Fairweather-Tait et al. 2005). To conclude, Caco-2 cell model approaches coupled with *in vitro* digestion are considered to be the most useful *in vitro* method to study iron availability from foods and to predict iron bioavailability *in vivo* although the magnitude of response is not comparable to humans (Fairweather-Tait et al. 2007). Iron solubility and dialyzability are considered less useful predictors of iron absorption, but can be used as a tool to understand factors that may affect iron absorption (Sandberg 2005). All three *in vitro* methods cannot be used alone for important decisions regarding iron (bio)fortification. Eventually, such decisions can only be taken based on information from human iron absorption studies (Fairweather-Tait et al. 2005).

Animal models in rats are useful tools to predict relative iron bioavailability of iron fortification compounds, especially if the compound is not yet released for the use in humans due to safety concerns (e.g. nanostructured iron compounds). The standard procedure, published by the Association of Official Analytical Chemist, includes iron-depleted anaemic rats which were fed with a test iron compound over a two week repletion phase. The ability of the iron compound to replete Hb in the rats is compared with the $FeSO_4$-control and reflects the RBV of the tested compound (Forbes et al. 1989). Animal models lack in sensitivity and as rats do not respond to enhancers and inhibitors of iron absorption in the same manner as humans, they are not an appropriate method to reflect bioavailability of dietary iron in humans (Reddy and Cook 1991; Fairweather-Tait et al. 2005).

3.5.2 Human isotope studies

Stable and radio iron isotopes are widely used for *in vivo* iron absorption studies investigating iron bioavailability from foods. The use of radioisotope in humans has a few advantages, but due to the exposure of subjects to ionizing radiation it is limited to non-pregnant, adult populations which are not the primary population groups vulnerable to ID. Therefore, stable isotope technique, which can be applied to all population groups of interest, is preferred over the radioisotope technique, and has been referred as the gold standard for measuring human iron absorption (International Atomic Energy Agency 2012). Comparison of absorption from stable and radioisotopes gave comparable results (Turnlund 1983; Barrett et al. 1994). On the downside, stable isotope technique is cost intensive and isotopic labelling of certain iron compound is very challenging. So far, it has not been possible to isotopically label iron elemental powders having the exact physical characteristics of their commercial counterparts. The differences in physical characteristics might influence their bioavailability (Hurrell et al. 2002). Another major disadvantage is that much larger quantities of stable isotopes than radioisotopes have to be added to the test meals and this change the molar ratios of iron with respects to the inhibitors and enhancers of iron absorption. While this is not a problem with added fortification iron, it is a problem to measure the absorption of native iron. In these situations, the addition of the stable isotope has to be split over multiple meals, making the studies much longer and more complicated.

The principle of stable isotope technique involves the oral administration of a single or multiple dose(s) of stable iron isotopes in form of an extrinsically or intrinsically labelled test meal. Iron absorption can then be estimated using one of the following three approaches: 1) faecal recovery, 2) plasma appearance or 3) erythrocyte iron incorporation. The majority of the studies use the erythrocyte iron incorporation approach which is more sensitive than the other approaches. Incorporation of iron isotopes into erythrocytes is measured in venous blood samples drawn 14 days after test meal administration using mass-spectrometry. Calculation of iron absorption is based on the fraction of absorbed iron incorporated into erythrocytes (for adults 80%) and on the blood volume estimated from weight, height and the iron content of blood calculated from Hb (International Atomic Energy Agency 2012).

As mentioned above, isotopes can be added to the test meal as either extrinsic or intrinsic tag. An extrinsic isotope tag is added to the test meal shortly before administration and directly mixed with the food matrix being studied. In the food matrix, the iron extrinsic tag is not integrated in the same way as the native iron found in the test meal, although it is expected to enter the common exchangeable iron pool of non-haem iron. Intrinsic labelling has the advantages that the iron tag is located in the same biological compartments as the other native iron (International Atomic Energy Agency 2012). Intrinsic labelling is obtained during the growth process of plants via stem injection, hydroponic culture or atmospheric labelling (Grusak 1997). It can be difficult and expensive, especially if plants do not accumulate much iron and/or if the edible parts of the plant are low in iron. Intrinsic labelling of some food sources, such as milk, has failed (Figueroacolon et al. 1989). In general, intrinsic and extrinsic labelling with iron has delivered comparable results in humans. This can be explained by the common exchangeable iron pool which occurs during digestion and in which the extrinsic labelling completely exchange with all the native non-haem food iron (Cook et al. 1972; Björn-Rasmussen et al. 1973; Consaul and Lee 1983). Therefore, successful extrinsic labelling requires a complete exchange between the native non-haem iron and the extrinsic label. This can be achieved by mixing the extrinsic tag properly with the food matrix and if possible by homogenizing the test meal components. In a study with un-milled unpolished rice, absorption from extrinsically labelled rice was significantly different from intrinsically labelled rice. This was not the case for rice flour. The authors suggested that the outer layer of the un-milled rice grains does not break up, and therefore prevent exchange of the intrinsic iron label during digestion (Björn-Rasmussen et al. 1973).

Another issue is the potential bias in single meal absorption studies. Some studies have revealed differences between iron absorption from single meals and multiple meals (Cook et al. 1991a; Tidehag et al. 1996). Multiple meal studies include the administration of several meals all containing a low dose of isotope label over a longer period of time. The study done by Tidehag et al. 1996 measured iron absorption from standardized meals which differed in iron concentration and which were first provided 3 days (3 times/day) without labelling and then labelled with radioisotopes for 2 days. The labelling in the breakfast meals was different from the other meals, and thus allowed a comparison of absorption between the single breakfast meals and the other meals. The study reported that the absorption from the breakfast meals were 50–80% higher than the average iron absorption from all meals during the same 2-day period. Authors suggested that the differences were either due to the lower iron intake from the breakfast meals in the morning (1/7 of total iron intake), causing a higher absorption, or due to the 12 hours overnight fast before breakfast, compared with only 4 hours without food before lunch and dinner. In the study done by Cook and colleagues, iron absorption from different single meals (low, moderate and high bioavailability) was compared with the absorption from complete diets (low, moderate and high bioavailability). Absorption from the complete diet was measured using a labelled bread roll which was consumed with 28 separate meals over a 2-week period. The influence of inhibitors (low bioavailability) and enhancers (high bioavailability) on iron absorption was much stronger in a single meal than in complete diets, whereas absorption from a single meal and complete diet with moderate bioavailability were similar. Iron absorption from the single meal with high bioavailability was 6-fold higher than the absorption from the single meal with low bioavailability, whereas the difference between complete diet with high and low bioavailability was only 2.5-fold. These findings indicate that single meal studies may overestimate the effects of enhancers and inhibitors compared to studies with complete diets. However, the complete diet in the study was not strictly controlled, and therefore variation in iron consumption between single meals and complete diet possibly contributed to the observed effects. In addition, the impact of inhibitors and enhancers in the complete diet was potentially reduced due to food residues in the stomach and duodenum of non-fasting subject compared to fasting subjects. It is also possible that the observed effect occurred due to a dilution of inhibitors and enhancers in the complex diet.

There are further studies supporting the assumption that single meals overestimate the effect of enhancer and inhibitors. The enhancing effect of AA on human iron absorption was much less pronounced in multiple meals than in single meals (Cook and Reddy 2001) and some long-term surveys have failed to demonstrate a correlation between iron status and the daily consumption of AA (Cook et al. 1984; Hunt et al. 1994), although others did demonstrate a correlation (Mao and Yao 1992) (see also chapter 3.3: Enhancing bioavailability of (bio)fortification iron). In another study, iron absorption from meals with low and high bioavailability was measured at baseline and after 10 weeks. In-between, the subjects, healthy young men, consumed the same meals in a 2-day circle. Adaptation decreased the difference in absorption between the low and high bioavailability meals from 8- to 4-fold after 10 weeks (Hunt and Roughead 2000). To conclude, single meal studies are a useful tool to identify factors influencing iron bioavailability, but are less appropriate to determine long-term iron bioavailability from complete diets. Multiple meal studies, which are more time-consuming and more expensive, are more accurate to estimate iron bioavailability from complete diets and from iron-biofortified foods.

4 IRON NUTRITION IN AREAS WITH ENDEMIC MALARIA

The findings of the Pemba trial (Sazawal et al. 2006) launched a vigorous debate about the safe administration of iron to infants, children, and pregnant women living in malaria endemic areas (Prentice 2008). In the Pemba trial, the incidence of all-cause morbidity and mortality was significantly increased in a group of more than 15'000 children receiving iron-folic acid supplementation compared to a control group (~8'000 children). In contrast, a subgroup of approximately 2'400 iron-deficient children who received the supplementation had fewer adverse events than the control children. Because Pemba is an area with very high malaria transmission and because a similar trial in Nepal showed no detrimental effects (Tielsch et al. 2006), the results have been ascribed to a harmful interaction between iron administration and malaria. It has to be emphasised that the adverse effects were only confined to iron-replete children suggesting that it was an excess of iron that caused the unfavourable interaction.

The key question that emerged after the Pemba trial was: "What are the potential mechanisms by which iron status could influence parasite invasion or growth, and thus the clinical sequelae of malarial infection?" The following chapters review some of the potential answers to this question. Furthermore, they give a general background on malaria, and they describe iron and malaria interactions and the safety of iron interventions in malarial areas.

4.1 Malaria

Malaria is an infectious disease caused by a protozoan of the genus *Plasmodium (P.)* transmitted into the human organism by the female

Anopheles mosquito. The four common *Plasmodium* species infecting humans are: *P. falciparum*, *P. vivax*, *P. malariae* and *P. ovale*. *P. falciparum* is the most deadly and responsible of the vast majority of morbidity and mortality (Crutcher and Hoffman 1996). Malaria is endemic in more than 100 countries causing over 200 million cases per year and about 1.6 million deaths, mainly in sub-Saharan Africa (Snow et al. 2005; WHO 2011c; Murray et al. 2012). Due to their less well-developed immune defence, infants and young children <5 years of age are particularly vulnerable to malarial infections (WHO 2011c) but also pregnant women and their unborn child are susceptible (Steketee et al. 2001). However, results are conflicting and a recently conducted study found that the number of deaths among children <5 years of age is probably not as high as a proportion of the total deaths as generally assumed. The study reported that in 2010 only 58% of the total deaths occurred in children <5 years of age (Murray et al. 2012), which is in strong contrast to the 86% reported by WHO (WHO 2011c).

4.1.1 The parasite's life cycle

The complex life cycle of *Plasmodium* parasite involves three phases. Phase 1 is the sexual stage of the parasite and occurs in the mosquito, whereas phase 2 and phase 3 are asexual stages taking place in the human host's liver and blood, respectively (Figure 11). During phase 1, when a mosquito draws a blood meal from a human already infected with *Plasmodium*, the male and female gametocytes of the *Plasmodium* find their way into the gut of the mosquito where they are able to fuse and where they are eventually converted into sporozoites (cell form of *P.* infectious for humans). The sporozoites are transferred to the salivary glands of the mosquito and introduced into the human skin when a mosquito draws blood from a human (Miller et al. 2002). While some of the sporozoites will be immediately destroyed by macrophages, others will reach the blood stream and move to the liver within a few hours where they invade hepatocytes (Silvie et al. 2008). In the liver (phase 2), the sporozoites undergo a process called schizogony in which each sporozoite forms a schizont (cell form of *P.* during replication of nucleus and organelles) containing tens of thousands of merozoites (cell form of *P.* infectious for human erythrocytes). After the rupture of the hepatocytes, these merozoites are released into the blood stream where they invade red blood cells (RBCs). Within the RBCs (phase 3), the merozoites replicate themselves asexually. Newly formed merozoites will be released by haemolysis of RBCs and invade other RBCs shortly after release (Miller et al. 2002). This phase is characterised by a rapid increase in parasite density and a drastic decrease in RBCs accompanied by fever in the host. A small proportion of merozoites is transformed into male or female gametocytes in phase 3, ready to be taken up by a mosquito for a new phase 1 (Figure 11) (Miller et al. 1994).

The reproductive cycle of *P. falciparum* and *P. vivax* merozoites in the RBCs lasts 48 hours. In the first 24 hours, the red blood cells infected with merozoites circulate freely but are rapidly cleared by the spleen. During the second 24 hours, the infected RBCs develop cup-shaped adhesive protrusions which allow them to adhere to the endothelial wall resulting in a sequestration in various organs (Miller et al. 2002). Human autopsy studies have shown sequestration of infected RBCs in varying degrees in the brain, intestine, heart, skin, kidney, liver, and lung. The sequestration of *P. falciparum*-infected RBCs is believed to be crucial for the disease pathogenesis and occurs in all *P. falciparum* infections, including those in asymptomatic individuals (Seydel et al. 2006).

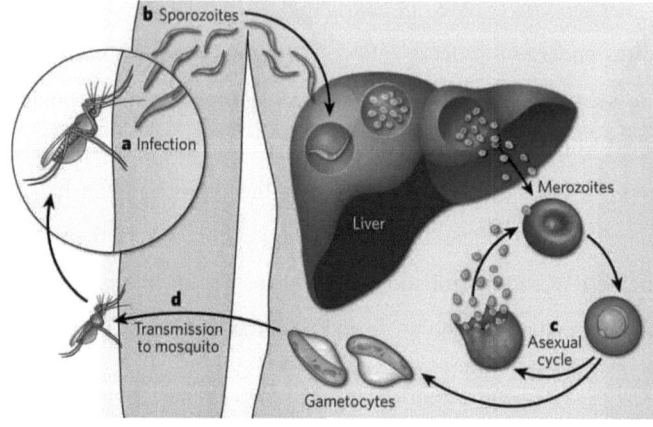

Figure 11: Life cycle of malaria including the phase in the mosquito and the liver and blood phases in the human host. a: An infected Anopheles mosquito injects sporozoites into a human host. b: Sporozoites are carried in the bloodstream to liver cells. c: Asexual reproduction cycle within red blood cells including the production of male and female gametocytes. d: Gametocytes are transmitted back to a mosquito where they fuse to form new sporozoites (Michalakis and Renaud 2009).

4.1.2 Asymptomatic vs. symptomatic malaria

Malaria parasitemia is often defined as asymptomatic or symptomatic which is not always straight forward because infected individuals might be in a pre-symptomatic period of parasitemia or the symptoms only occurred for a very short period without being noticed. In general, asymptomatic malaria is defined as the presence of *Plasmodium* parasites in peripheral thick blood smears without fever (body temperature <37.5°C) and without malaria-related symptoms such as fatigue or malaise (Laishram et al. 2012). However, this definition does not include the so-called sub-patent malaria parasitemia (undetectable by microscopy) where the parasites appear only in the liver. A study using a polymerase chain reaction reported high prevalence of sub-patent malaria parasitemia in an endemic area in Senegal (Bottius et al. 1996).

In endemic areas, symptomatic febrile malaria is believed to be primarily a disease of young children <5 years of age who have little acquired immunity (Hurrell 2010a). Asymptomatic malaria is a consequence of constant exposure including repeated febrile malarial episodes which then result in the development of immunity so that older children, adolescents and adults may have chronic parasitemia without symptoms. The prevalence and density of parasitemia, as well as the level of infection-induced inflammation depend on age, pregnancy status and previous exposure to infection (Verhoef 2010). High prevalence of asymptomatic malaria was previously reported in pregnant women (58–80%) (Onyenekwe et al. 2002; Nwagha et al. 2009), in school children (40–96%) (Githeko et al. 1993; Kimbi et al. 2005), or in adult males and females (24–44%) (Eke et al. 2006). It is assumed that in some parts of Africa at any point of time; almost 100% of the children have asexual stage of malaria parasitemia in their blood (Ghosh 2007). Prevalence of asymptomatic and symptomatic malaria varies among rainy and dry season (Onyenekwe et al. 2002) and is expected to be low in areas with low malaria transmission (Zoghi et al. 2012).

The acquired immunity is sufficient to control but not to prevent malaria infection. It is slowly acquired and quickly lost in infection-free conditions. Without treatment, asymptomatic *Plasmodium* infections may persist for months (Verhoef 2010). Individuals carrying asymptomatic malaria provide an essential reservoir of parasites, contributing to the persistence of malaria transmission (Laishram et al. 2012). There is evidence that asymptomatic malaria induces the same inflammation patterns as symptomatic malaria but to a lower magnitude (de Mast et al. 2009; de Mast et al. 2010) (see also chapter 4.2: Interaction of malaria with human iron homeostasis). After malarial treatment, reinfection can occur rapidly because individuals in endemic areas receive 3–4 infectious mosquito bites per night. Reinfections might be asymptomatic or symptomatic (Verhoef 2010).

4.2 Interaction of malaria with human iron homeostasis

Malaria affects hematologic status of its host, but how does the hematologic and iron status of the host affect malaria and the susceptibility to it? In addition to the Pemba study, other studies have also investigated the impact of iron status on malaria susceptibility. ID has been reported to protect against severe *P. falciparum* malaria and death in young children (Gwamaka et al. 2012), to decrease susceptibility to *P. falciparum* infection in pregnant women (Kabyemela et al. 2008) and to reduce the risk of placental malaria (Senga et al. 2011). Furthermore, susceptibility to malaria was increased in children with high SF concentrations (Snow et al. 1991; Nyakeriga et al. 2004) or in infants with higher Hb concentrations (Oppenheimer et al. 1986). In contrast to all these studies, one study reported that anaemic infants were marginally more susceptible to malaria than non-anaemic infants (Prentice et al. 2007). All these studies reveal certain interactions between malaria and its hosts iron homeostasis. The following chapters give a brief overview on human iron homeostasis before having a closer look into these interactions.

4.2.1 Human iron homeostasis

In non-transfused individuals, iron moves into the body exclusively through the diet. Due to the lack of regulated iron excretion through the liver or kidney, iron balance is primarily controlled at the level of intestinal absorption (Andrews 2008). About 1–2 mg iron from the diet are taken up daily by the enterocytes in the proximal part of the duodenum. A big part of this iron is used for erythropoiesis which takes place in the bone marrow. Erythropoiesis is not only based on dietary iron but as well on the daily recycling of 20 to 25 mg iron from aged RBCs. Recycling is done by macrophages which phagocytize aged RBCs in the spleen, in the bone marrow and in the liver (Figure 12) (Hentze et al. 2004).

Dietary iron is mostly non-haem iron which enters the gut usually in the Fe^{3+} form (Hulten et al. 1995) (see also chapter 2: Dietary Factors Affecting Iron Bioavailability from Sorghum and Millet Foods). To enter the enterocytes, Fe^{3+} iron has to be reduced to Fe^{2+} iron, either chemically or by a ferrireductase such as duodenal cytochrome B. Enterocytes absorb Fe^{2+} iron through a carrier protein called divalent metal transporter 1. Iron export from both, enterocytes and macrophages, is mediated by ferroportin, the only known iron export protein. Once iron is exported it is oxidized into Fe^{3+} iron and bound to TF, a serum transport protein with two binding sites for Fe^{3+} iron (Andrews 2008). TF transports iron to the bone marrow, the liver or to other cells and tissues (Figure 12) (Hentze et al. 2004). Iron that is bound to ligands other than TF is named non-transferrin bound iron (NTBI) and is suspected to influence malarial infections (Hurrell 2010a). Discussion on the interaction of NTBI and malaria follows in a later chapter.

Iron uptake and recycling is controlled by hepcidin, an amino acid peptide mainly synthesised in the liver. Hepcidin binds to ferroportin of enterocytes and macrophages causing the internalization and degradation of ferroportin. By doing this, hepcidin simultaneously blocks dietary iron absorption and prevents iron recycling from aged RBCs (Figure 12). Therefore, elevated hepcidin concentrations restrict iron availability to the erythron, while lowered hepcidin concentrations result in increased iron absorption from the diet and in a more efficacious recycling of iron via macrophages. Hepcidin is homeostatically up-regulated in high iron status (Nemeth and Ganz 2009) but as well during inflammation (Nemeth et al. 2004). Up-regulated hepcidin during malarial infections (Howard et al. 2007) plays a key role in the interaction of malaria with iron homeostasis (Spottiswoode et al. 2012) and is discussed in more detail in the following sections.

4.2.2 Malarial anaemia

In populations where malaria infections are common, the prevalence of anaemia is high (Crawley 2004). Morbidity and mortality from malaria is associated with anaemia which is often severe with Hb concentrations of less than 50 g/L (Kurtzhals et al. 1997). The exact mechanism leading to anaemia in malarial infections is not fully understood, but the aetiology is most likely multifactorial. The major contributors are the clearance of infected and uninfected RBCs by the macrophages in the spleen, the rupture of infected cells in the circulation during merozoites reproduction, and an inadequate erythroid response due to inflammation and various parasite products (Lamikanra et al. 2007). An innate immune response to parasitic infections, in which acute phase reactants restrict parasitic growth by withholding iron in the macrophages, may also contributes to malarial anaemia (Wander et al. 2009). The reason for this so-called anaemia of infection is the increase in hepcidin concentrations in response to inflammation (Figure 12) (Ganz 2007).

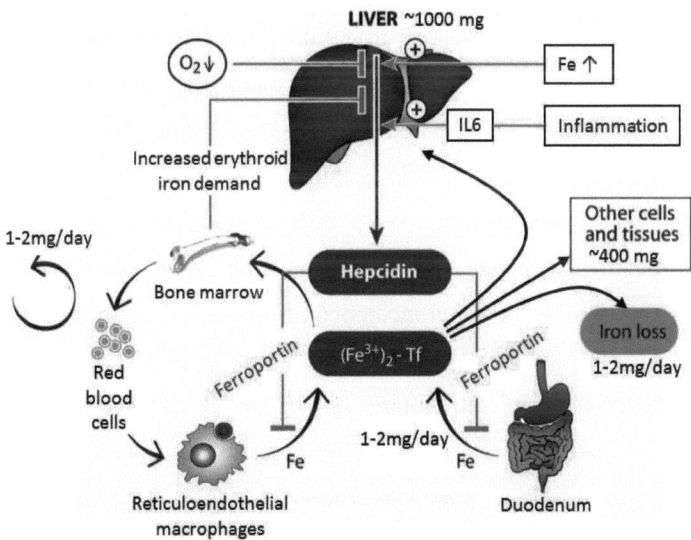

Figure 12: Systemic iron homeostasis and the role of hepcidin. Approximate values for the iron content of different organs and tissues in humans are stated, and the approximate daily fluxes of iron are also indicated. Hepcidin diminishes iron release from reticuloendothelial macrophages and duodenal enterocytes by causing ferroportin's internalization and degradation. Hepcidin expression is regulated by iron levels, inflammatory stimuli, the erythroid iron demand, and hypoxia (modified from Hentze et al. 2004). Tf, transferrin; Fe, iron; IL-6, interleukin-6.

4.2.3 Influence of malaria on iron absorption

The acute phase response to malaria (O'Donnell et al. 2009) increases cytokine concentrations, such as interleukin-6 (Lyke et al. 2004), which can stimulate hepcidin expression (Ganz 2007). Elevated concentrations of hepcidin during malarial infection are assumed to impair iron absorption by blocking iron efflux from enterocytes. A study investigating incorporation of oral iron into RBCs and the efficacy of iron supplementation in young Gambian children immediately after treatment of febrile malaria supports this assumption. In this study, the RBC incorporation of an oral iron dose was significantly lower in anaemic children, who consumed the oral iron 1 day after they received malarial treatment, than in control children with anaemia but without malarial episode before iron administration (Doherty et al. 2008). The lower iron incorporation in the post-malaria children was likely due to elevated hepcidin concentrations which were not yet normalized shortly after treatment and which blocked the iron export from the enterocytes (Hurrell 2010a). Two weeks later, iron incorporation was tested again. This time, incorporation of oral iron into RBCs in the post-malaria group was no longer different from the control group which showed the same incorporation rate as 2 weeks earlier. This suggests that iron incorporation and likely iron absorption were normalized in the post-malaria children due to the normalization of hepcidin concentrations 2 weeks after the malaria episode. However, interpretation of these results is complicated by the fact that the iron dose and the AA:Fe were not the same on the first administration and 2 weeks later.

As a secondary outcome, the study also investigated the repletion of Hb in the two groups of children. Despite the lower iron absorption, the post-malarial anaemic children responded much better to iron supplementation than the non-malarial anaemic children. As erythropoiesis is mainly based on iron from aged RBCs, the authors argued that the release of iron from macrophages after treatment was responsible for the rapid Hb repletion in the post-malarial group. During the malarial episode, the iron efflux from macrophages is blocked which leads to an increased store of recyclable iron from the phagocytosis of infected as well as aged RBCs. After malarial treatment the iron is released and can be used for erythropoeitic needs during the first two weeks of recovery. This suggest that iron released from the macrophages is sufficient to immediately improve Hb concentrations, and therefore it is maybe better to introduce iron supplementation later when iron absorption is already normalized (Hurrell 2010a).

4.2.4 Malaria and non-transferrin bound iron

Increased plasma NTBI concentrations have been detected in subjects consuming oral iron supplementations (Hutchinson et al. 2004; Dresow et al. 2008) and are suggested to cause the harmful effects in iron-supplemented subjects living in malaria endemic areas by interacting with the liver stage and possibly also with the blood stage of the malaria parasites (Hurrell 2010a). NTBI is much faster taken up by hepatocytes than TF-bound iron (Brissot et al. 1985). Hurrell 2010 suggested that NTBI entering the liver might facilitate the penetration of hepatocytes by sporozoites. This is consistent with an animal study in which iron overload increased hepatic development of *Plasmodium* parasitemia (Goma et al. 1996) and with a study in which growth of *Plasmodium* parasites in hepatocytes cultures was inhibited by iron chelators (Loyevsky et al. 1999).

The high rate of deoxyribonucleic acid synthesis during the blood phase of *Plasmodium* is catalysed by a ribonucleotide reductase which requires iron (Gordeuk et al. 1992). It is suggested that the parasite derives its iron from within the infected RBCs (Hershko 2007), but the mechanism by which it obtains the iron is not entirely clear. It is likely that *Plasmodium* obtains the iron from a labile, chelatable intracellular iron pool most probably derived from haem catabolism and not directly from the Hb (Hershko and Peto 1988). Iron bound to TF is not taken up by infected RBCs (Peto and Thompson 1986) and although an *in vitro* study showed NTBI uptake by infected and uninfected RBCs (Sanchez-Lopez and Haldar 1992), it seems unlikely that NBTI is able to enter RBCs (Hurrell 2010a). Therefore, it is improbable that NBTI directly affects development of *Plasmodium* during the blood phase.

It is, however, likely that NTBI influences the sequestration of malaria-infected RBCs. During the reproduction cycle, *Plasmodium* merozoites cover the surface of RBCs with small cup-shaped protrusions containing a membrane protein-1 (PfEMP-1) (Horrocks et al. 2005). PfEMP-1 binds to various host ligands, such as intercellular adhesion molecule-1 (ICAM-1) and vascular adhesion molecule-1 (VCAM-1), enabling the adherence to blood vessels walls. This sequestration from the circulation avoids the removal of infected RBCs by the spleen (Miller et al. 2002). In malarial episodes, ICAM-1 is up-regulated in response to pro-inflammatory mediators and associated with cerebral malaria where it facilitates sequestration of RBCs in the microvasculature of the brain (Turner et al. 1998). NTBI has been reported to increase concentrations of adhesion molecules (ICAM-1, VCAM-1) in humans, mostly likely by an inflammatory process caused through free radical generation (Kartikasari et al. 2006). Such an increase is expected to facilitate sequestration of infected RBCs and contribute to malaria morbidity and mortality (Hurrell 2010a).

Sequestration of infected RBCs is particularly high in the small capillaries of the intestine (Seydel et al. 2006). Increased NTBI concentration might augment these sequestration by the mechanism mentioned above. It is likely that such a sequestration and additional pro-inflammatory and prooxidant stress caused by supplemental iron lead to a breakdown of gut barrier resulting in the systemic invasion of numerous pathogenic bacteria. There is a clear association between malaria and non-typhoidal *Salmonella* disease in African children (Graham 2010). Growth of *Salmonella* and other pathogenic enterobacteria, such as *E. coli* and *shigella*, benefit from unabsorbed supplemental iron doses, and thus impede the competitive inhibition of pathogenic bacteria by beneficial bacteria, such as lactobacilli and bifidobacteria, which do not need iron for their growth. This assumption is supported by a recent RCT among Ivoirian school children who received iron-fortified biscuits for 6 months (20 mg electrolytic iron, 4 times a week). Compared to the control group, the children receiving iron had a more unfavourable ratio of fecal enterobacteria to bifidobacteria and lactobacilli, and a significant increase in the gut inflammation marker calprotectin (Zimmermann et al. 2010). Furthermore, the pathogens, which invade the systemic circulation, possibly also benefit from increased NTBI concentrations (Hurrell 2010a).

The magnitude of the influence of NTBI on malaria infections is related to the degree of NTBI increase in the systemic circulation, which again is linked to the amount of absorbed iron during digestion. The increase of NTBI concentrations will be lower under certain conditions such as when 1) the iron dose is relatively low, 2) iron is administered with food which decreases absorption by as much as two thirds (Hurrell 2010a), 3) the subject is iron replete, 4) there is clinical or subclinical infection or inflammation which decreases absorption (Dewey and Baldiviez 2012), 5) the iron form (e.g. NaFeEDTA) per se has slower plasma appearance (Schumann et al. 2012).

4.3 Safety of iron interventions in malarial areas

As a result to the Pemba trial, WHO radically changed their recommendations for iron supplementation of children <2 years of age living in malaria endemic areas. The universal iron supplementation was replaced by a targeted supplementation for only those with detected ID where a screening system for ID is available and for only those with clinical symptoms of severe anaemia where screening is not available (WHO 2007). In addition to the Pemba trial, other studies also found an increased malaria risk when administering iron supplements. In a controlled study with anaemic Somali nomads 11–60 years old, iron supplementation led to a significant increase in Hb in the intervention group but also to a significant increase in clinical malaria attacks (Murray et al. 1978). In Peruvian children over 5 years of age, malaria episodes (*P. vivax*) significantly increased by 49% in the iron-supplemented group (Richard et al. 2006). Another study with anaemic Gambian children, 6–60 months of age, reported significantly improved iron status with iron supplementation but also significantly increased fever episodes which were associated with malaria parasitemia (Smith et al. 1989). Several further studies showed a trend of increased malaria risk in iron-supplemented groups compared to placebo groups. These studies reported a non-significant increase in percentage of infants or children with parasitemia (Harvey et al. 1989; Chippaux et al. 1991; Verhoef et al. 2002; Desai et al. 2003), an increase in percentage of children with parasitemia >3000 (Berger et al. 2000) or an increased treatment failure of malaria (Nwanyanwu et al. 1996). In contrast, six studies described no effect of iron supplementation on malaria risk. These studies were done in infants and children with high anaemia prevalence (Lawless et al. 1994;

van Hensbroek et al. 1995; van den Hombergh et al. 1996; Menendez et al. 1997; Mebrahtu et al. 2004) and in pregnant women (Menendez et al. 1994). It has to be stressed that interpretation and comparison of all these studies is complicated due to the fact that they differ in the iron dose, iron status, the percentage of malaria at baseline and in the administration of malaria treatment. However, one outstanding difference between the studies, showing increased malaria risk and those with no effect, is access to health care facilities and active follow-up and treatment of malaria incident cases. This implies that concurrent effective treatment of malaria infections may reduce the potential adverse effect of iron supplementation (Prentice et al. 2007). This is supported by the two recent Cochrane meta-analyses (Ojukwu et al. 2009; Okebe et al. 2011) which lead to the conclusion that oral iron supplements given to children living in malaria endemic areas do not increase the risk of clinical malaria or death if malaria surveillance is provided simultaneously and if treatment services are available. In other words, if malaria surveillance and health care is inadequate, as it is the case in many developing countries, iron supplementation to young children in areas with endemic malaria can cause increased malarial morbidity and mortality (Hurrell 2011).

Taking into account the now strong evidence that iron supplementation in malaria endemic areas can lead to adverse effects in young children under certain circumstances and that the mechanisms for these adverse effects are assumed to be linked to NTBI, the appropriate approach for limiting risk is the shift towards food-based interventions e.g. adapted fortification with decreased iron doses since these interventions are not expected to increase NTBI concentrations. In addition, the iron interventions should always be coordinated with malaria control interventions (Stoltzfus 2012). Indeed, studies on *in-home* fortification, including malaria incidence as a primary or secondary outcome, reported no increased risk for adverse effects (Adu-Afarwuah et al. 2007; Zlotkin et al. 2011; Suchdev et al. 2012). However, the sample size in these studies was not large and considered too small to adequately assess potential adverse effects. Further research to prove the current evidence that (*in-home*) fortification is safe in malaria endemic areas would therefore be valuable (Dewey and Baldiviez 2012) as well as further research on the interactions between malaria and iron homeostasis (Prentice and Cox 2012).

An advantage of (*in-home*) fortification is that screening for ID, as recommended by WHO, would not be necessary. Screening for ID seems to be an unrealistic and unpractical option in multiple ways, particularly in developing countries. First, it is difficult to decide on which indicator the screening should be based. Interpretation of many indicators and their cut-offs is problematic because they are affected by infections and lead exposure (Stoltzfus 2012). Second, cost concerns have to be considered. Costs can be reduced with better and cheaper technologies, but at the moment screening for ID is only possible in expensive well-equipped clinical settings. Cheaper field-friendly methods to diagnose ID in children are still under development. The most acceptable and potentially least expensive approaches to assess hematologic and iron status would be non-invasive methods, i.e. methods not requiring the extraction of blood. Current science is far away from such approaches but first skin-probe Hb devices are being developed and tested (Crowley et al. 2012). Third, iron interventions to targeted children screened for ID is a treatment strategy without prevention aspects. Considering that ID might cause irreversible adverse effects during child development, prevention strategies, such as iron fortification, should have priority (Stoltzfus 2012).

4.4 Impact of malaria on the efficacy of iron fortification

It is expected that both, acute febrile and asymptomatic malaria, have a negative impact on human iron absorption. Regarding long-term impact, it is assumed that asymptomatic malaria plays a more important role as it affects vulnerable population groups for most of the year (Nwagha et al. 2009) (see also chapter 4.1.2: Asymptomatic vs. symptomatic malaria). In contrast, acute febrile malaria episode are limited to a few days a year (Hurrell 2010a). Asymptomatic malaria triggers some low-grade inflammation including hepcidin increase which potentially blunts the efficacy of iron fortification programs. However, at the present time there is no strong evidence for that. Some evidence comes from two fortification trials using electrolytic iron. They were conducted in malaria endemic areas in Kenyan (Andang'o et al. 2007) and Ivorian (Rohner et al. 2010) children and have reported no efficacy. This is in contrast to two studies that reported efficacy when using the same iron compound in non-endemic malarial areas in Chinese school children (Sun et al. 2007) and Thai women (Zimmermann et al. 2005b). However, it is also likely that the blunted effect observed in the studies conducted in Africa is due to a poorer production of gastric acid in malnourished children, so that electrolytic iron is less well absorbed because it has not been dissolved properly in the gastric juice (Hurrell 2010a). The study done in Kenya also used NaFeEDTA in one arm and reported efficacy for that compound. A similar study done in a non-malaria transmission area in India found very similar results as the Kenya study suggesting that there is no blunting by malaria in the Kenya study. In addition, no evidence of blunting was found when comparing two studies using FePP-fortified salt. One study was done in Moroccan children with no malaria infections (Zimmermann et al. 2004b) and the other in Ivorian children with high prevalence of malaria and other infections (Wegmuller et al. 2006). Both studies reported good efficacy for the iron-fortified salt.

Comparison of efficacy studies on iron fortification done in malarial areas with studies conducted in non-malarial areas has to be done with caution because the iron dose, baseline characteristics of the subjects and the fortified vehicle differ between the studies (Hurrell 2010a). Further research is needed to make clearer statement about the efficacy of iron fortification in malaria endemic areas. A study with an intervention group receiving iron-fortified foods and malaria prophylaxis compared to an intervention group receiving only iron-fortified foods would deliver interesting results, but is perhaps difficult to apply from an ethical point of view.

REFERENCES

Abdalla, A. A., A. H. El Tinay, B. E. Mohamed and A. H. Abdalla (1998). Proximate composition, starch, phytate and mineral contents of 10 pearl millet genotypes. *Food Chem* **63**(2): 243-246.

Abdelrahaman, S. M., H. B. Elmaki, W. H. Idris, A. B. Hassan, E. E. Babiker and A. H. El Tinay (2007). Antinutritional factor content and hydrochloric acid extractability of minerals in pearl millet cultivars as affected by germination. *Int J Food Sci Nutr* **58**(1): 6-17.

Abdelrahman, A., R. C. Hoseney and E. Varrianomarston (1984). The Proportions and Chemical-Compositions of Hand-Dissected Anatomical Parts of Pearl-Millet. *J Cereal Sci* **2**(2): 127-133.

Abebe, Y., A. Bogale, K. Hambidge, B. Stoecker, K. Bailey and R. Gibson (2007). Phytate, zinc, iron and calcium content of selected raw and prepared foods consumed in rural Sidama, Southern Ethiopia, and implications for bioavailability. *J Food Compos Anal* **20**(3-4): 161-168.

Abizari, A. R., D. Moretti, S. Schuth, M. B. Zimmermann, M. Armar-Klemesu and I. D. Brouwer (2012a). Phytic Acid-to-Iron Molar Ratio Rather than Polyphenol Concentration Determines Iron Bioavailability in Whole-Cowpea Meal among Young Women. *J Nutr* **142**(11): 1950-1955.

Abizari, A. R., D. Moretti, M. B. Zimmermann, M. Armar-Klemesu and I. D. Brouwer (2012b). Whole Cowpea Meal Fortified with NaFeEDTA Reduces Iron Deficiency among Ghanaian School Children in a Malaria Endemic Area. *J Nutr* **142**(10): 1836-1842.

Abrams, S. A., A. Mushi, D. C. Hilmers, I. J. Griffin, P. Davila and L. Allen (2003). A multinutrient-fortified beverage enhances the nutritional status of children in Botswana. *J Nutr* **133**(6): 1834-1840.

Adu-Afarwuah, S., A. Lartey, K. H. Brown, S. Zlotkin, A. Briend and K. G. Dewey (2007). Randomized comparison of 3 types of micronutrient supplements for home fortification of complementary foods in Ghana: effects on growth and motor development. *Am J Clin Nutr* **86**(2): 412-420.

Adu-Afarwuah, S., A. Lartey, K. H. Brown, S. Zlotkin, A. Briend and K. G. Dewey (2008). Home fortification of complementary foods with micronutrient supplements is well accepted and has positive effects on infant iron status in Ghana. *Am J Clin Nutr* **87**(4): 929-938.

Afify, A. M. R., H. S. El-Beltagi, S. M. Abd El-Salam and A. A. Omran (2011). Bioavailability of Iron, Zinc, Phytate and Phytase Activity during Soaking and Germination of White Sorghum Varieties. *PLoS One* **6**(10).

Aherne, S. A. and N. M. O'Brien (2002). Dietary flavonols: Chemistry, food content, and metabolism. *Nutrition* **18**(1): 75-81.

Allen, L. H. (2002). Advantages and limitations of iron amino acid chelates as iron fortificants. *Nutr Rev* **60**(7): S18-S21.

Almgard, G. (1963). High content of iron in teff, Eragrostis Abysinia Link; and some other crop species from Ethiopia - a result of contamination. *Ann Afric Coll* **29**: 215-220.

Amarowicz, R., R. Carle, G. Dongowski, A. Durazzo, R. Galensa, D. Kammerer, . . . M. K. Piskula (2009). Influence of postharvest processing and storage on the content of phenolic acids and flavonoids in foods. *Mol Nutr Food Res* **53**: S151-S183.

Andang'o, P. E. A., S. J. M. Osendarp, R. Ayah, C. E. West, D. L. Mwaniki, C. A. De Wolf, . . . H. Verhoef (2007). Efficacy of iron-fortified whole maize flour on iron status of schoolchildren in Kenya: a randomised controlled trial. *Lancet* **369**(9575): 1799-1806.

Andersson, M., W. Theis, M. B. Zimmermann, J. T. Foman, M. Jakel, G. S. M. J. E. Duchateau, . . . R. F. Hurrell (2010). Random serial sampling to evaluate efficacy of iron fortification a randomized controlled trial of margarine fortification with ferric pyrophosphate or sodium iron edetate. *Am J Clin Nutr* **92**(5): 1094-1104.

Andjelkovic, M., J. Van Camp, B. De Meulenaer, G. Depaemelaere, C. Socaciu, M. Verloo and R. Verhe (2006). Iron-chelation properties of phenolic acids bearing catechol and galloyl groups. *Food Chem* **98**(1): 23-31.

Andrews, N. C. (2008). Forging a field: the golden age of iron biology. *Blood* **112**(2): 219-230.

Angeles-Agdeppa, I., M. V. Capanzana, C. V. Barba, R. F. Florentino and K. Takanashi (2008). Efficacy of iron-fortified rice in reducing anemia among schoolchildren in the Philippines. *Int J Vitam Nutr Res* **78**(2): 74-86.

Anglani, C. (1998). Sorghum for human food - A review. *Plant Food Hum Nutr* **52**(1): 85-95.

Aniszewski, T., R. Lieberei and K. Gulewicz (2008). Research on Catecholases, Laccases and Cresolases in Plants. Recent Progress and Future Needs. *Acta Biol Cracov Bot* **50**(1): 7-18.

Arcanjo, F. P., O. M. Amancio, J. A. Braga and V. de Paula Teixeira Pinto (2010). Randomized controlled trial of iron-fortified drinking water in preschool children. *J Am Coll Nutr* **29**(2): 122-129.

Archana, S. Sehgal and A. Kawatra (1998). Reduction of polyphenol and phytic acid content of pearl millet grains by malting and blanching. *Plant Food Hum Nutr* **53**(2): 93-98.

Ash, D. M., S. R. Tatala, E. A. Frongillo, G. D. Ndossi and M. C. Latham (2003). Randomized efficacy trial of a micronutrient-fortified beverage in primary school children in Tanzania. *Am J Clin Nutr* **77**(4): 891-898.

Ashok Kumar, A., B. V. S. Reddy, R. B., P. Sanjana Reddy, K. L. Sahrawat and H. D. Upadhyaya (2009). Variability for grain iron and zinc content in pearl millet hybrids. *Journal of SAT agricultural research* **7**: 1-4.

Awika, J. M. (2011) Sorghum flavanoids: unusual compounds with promising implications for health. In *Advances in cereal science: implications to food processing and health promotion*, pp. 171-200. Washington DC: American Chemical Society.

Awika, J. M., L. Dykes, L. Gu, L. W. Rooney and R. L. Prior (2003a). Processing of sorghum (Sorghum bicolor) and sorghum products alters procyanidin oligomer and polymer distribution and content. *J Agr Food Chem* **51**(18): 5516-5521.

Awika, J. M., L. Dykes, L. W. Gu, L. W. Rooney and R. L. Prior (2003b). Processing of sorghum (Sorghum bicolor) and sorghum products alters procyanidin oligomer and polymer distribution and content. *Journal of Agricultural and Food Chemistry* **51**(18): 5516-5521.

Awika, J. M. and L. W. Rooney (2004). Sorghum phytochemicals and their potential impact on human health. *Phytochemistry* **65**(9): 1199-1221.

Awika, J. M., L. W. Rooney and R. D. Waniska (2004). Properties of 3-deoxyanthocyanins from sorghum. *J Agr Food Chem* **52**(14): 4388-4394.

Awika, J. M., L. W. Rooney and R. D. Waniska (2005). Anthocyanins from black sorghum and their antioxidant properties. *Food Chem* **90**(1-2): 293-301.

Azeke, M. A., S. J. Egielewa, M. U. Eigbogbo and I. G. Ihimire (2011). Effect of germination on the phytase activity, phytate and total phosphorus contents of rice (Oryza sativa), maize (Zea mays), millet (Panicum miliaceum), sorghum (Sorghum bicolor) and wheat (Triticum aestivum). *J Food Sci Tech Mys* **48**(6): 724-729.

Backstrand, J. R., L. H. Allen, A. K. Black, M. de Mata and G. H. Pelto (2002). Diet and iron status of nonpregnant women in rural Central Mexico. *Am J Clin Nutr* **76**(1): 156-164.

Badau, M. H., I. Nkama and I. A. Jideani (2005). Phytic acid content and hydrochloric acid extractability of minerals in pearl millet as affected by germination time and cultivar. *Food Chem* **92**(3): 425-435.

Badi, S., B. Pedersen, L. Monowar and B. O. Eggum (1990). The nutritive value of new and traditional sorghum and millet foods from Sudan. *Plant Foods Hum Nutr* **40**(1): 5-19.

Ballot, D. E., A. P. Macphail, T. H. Bothwell, M. Gillooly and F. G. Mayet (1989). Fortification of Curry Powder with NaFe(III)EDTA in an Iron-Deficient Population - Report of a Controlled Iron-Fortification Trial. *Am J Clin Nutr* **49**(1): 162-169.

Baltussen, R., C. Knai and M. Sharan (2004). Iron fortification and iron supplementation are cost-effective interventions to reduce iron deficiency in four subregions of the world. *J Nutr* **134**(10): 2678-2684.

Barahona, R., C. E. Lascano, R. Cochran, J. Morrill and E. C. Titgemeyer (1997). Intake, digestion, and nitrogen utilization by sheep fed tropical legumes with contrasting tannin concentration and astringency. *J Anim Sci* **75**(6): 1633-1640.

Barrett, J. F. R., P. G. Whittaker, J. D. Fenwick, J. G. Williams and T. Lind (1994). Comparison of Stable Isotopes and Radioisotopes in the Measurement of Iron-Absorption in Healthy Women. *Clin Sci* **87**(1): 91-95.

Beck, K., C. A. Conlon, R. Kruger, J. Coad and W. Stonehouse (2011). Gold kiwifruit consumed with an iron-fortified breakfast cereal meal improves iron status in women with low iron stores: a 16-week randomised controlled trial. *Br J Nutr* **105**(1): 101-109.

Beebe, S., A. V. Gonzalez and J. Rengifo (2000). Research on trace minerals in the common bean. *Food Nutr Bull* **21**(1): 387-391.

Beecher, G. R. (2003). Overview of dietary flavonoids: Nomenclature, occurrence and intake. *J Nutr* **133**(10): 3248s-3254s.

Beinner, M. A., G. Velasquez-Melendez, M. C. Pessoa and T. Greiner (2010). Iron-fortified rice is as efficacious as supplemental iron drops in infants and young children. *J Nutr* **140**(1): 49-53.

Belton, P. S. and J. R. N. Taylor (2002) *Pseudocereals and less common cereals : grain properties and utilization potential*. Berlin ; New York: Springer.

Berger, J., J. L. Dyck, P. Galan, A. Aplogan, D. Schneider, P. Traissac and S. Hercberg (2000). Effect of daily iron supplementation on iron status, cell-mediated immunity, and incidence of infections in 6-36 month old Togolese children. *Eur J Clin Nutr* **54**(1): 29-35.

Besrat, A., A. Admasu and M. Ogbai (1980). Critical sutdy of the iron content of tef (eragrostis tef). *Ethiopian Med J* **18**: 45-52.

Beta, T., L. W. Rooney, L. T. Marovatsanga and J. R. N. Taylor (1999). Phenolic compounds and kernel characteristics of Zimbabwean sorghums. *J Sci Food Agric* **79**(7): 1003-1010.

Biebinger, R., M. B. Zimmermann, S. N. Al-Hooti, N. Al-Hamed, E. Al-Salem, T. Zafar, . . . R. F. Hurrell (2009). Efficacy of wheat-based biscuits fortified with microcapsules containing ferrous sulfate and potassium iodate or a new hydrogen-reduced elemental iron: a randomised, double-blind, controlled trial in Kuwaiti women. *Brit J Nutr* **102**(9): 1362-1369.

Björn-Rasmussen, E. and L. Hallberg (1979). Effect of animal proteins on the absorption of food iron in man. *Nutr Metab* **23**(3): 192-202.

Björn-Rasmussen, E., L. Hallberg and R. B. Walker (1973). Food Iron-Absorption in Man .2. Isotopic-Exchange of Iron between Labeled Foods and between a Food and an Iron Salt. *Am J Clin Nutr* **26**(12): 1311-1319.

Blandino, A., M. E. Al-Aseeri, S. S. Pandiella, D. Cantero and C. Webb (2003). Cereal-based fermented foods and beverages. *Food Res Int* **36**(6): 527-543.

Blessin, C. W., C. H. Vanetten and R. Wiebe (1958). Carotenoid Content of the Grain from Yellow Endosperm-Type Sorghums. *Cereal Chem* **35**(5): 359-365.

Boech, S. B., M. Hansen, K. Bukhave, M. Jensen, S. S. Sorensen, L. Kristensen, . . . B. Sandstrom (2003). Nonheme-iron absorption from a phytate-rich meal is increased by the addition of small amounts of pork meat. *Am J Clin Nutr* **77**(1): 173-179.

Bohm, H., H. Boeing, J. Hempel, B. Raab and A. Kroke (1998). Flavonols, flavones and anthocyanins as native antioxidants of food and their possible role in the prevention of chronic diseases. *Z Ernahrungswiss* **37**(2): 147-163.

Bohn, T., L. Davidsson, T. Walczyk and R. F. Hurrell (2004). Phytic acid added to white-wheat bread inhibits fractional apparent magnesium absorption in humans. *Am J Clin Nutr* **79**(3): 418-423.

Boren, B. and R. D. Waniska (1992). Sorghum seed color as an indicator of tannin content. *J Appl Poultry Res* **1**: 117-121.

Borg, S., H. Brinch-Pedersen, B. Tauris and P. B. Holm (2009). Iron transport, deposition and bioavailability in the wheat and barley grain. *Plant Soil* **325**(1-2): 15-24.

Bothwell, T. H. and A. P. MacPhail (2004). The potential role of NaFeEDTA as an iron fortificant. *Int J Vitam Nutr Res* **74**(6): 421-434.

Bottius, E., A. Guanzirolli, J. F. Trape, C. Rogier, L. Konate and P. Druilhe (1996). Malaria: even more chronic in nature than previously thought; evidence for subpatent parasitaemia detectable by the polymerase chain reaction. *Trans R Soc Trop Med Hyg* **90**(1): 15-19.

Bouis, H. E., C. Hotz, B. McClafferty, J. V. Meenakshi and W. H. Pfeiffer (2011). Biofortification: A new tool to reduce micronutrient malnutrition. *Food Nutr Bull* **32**(1): S31-S40.

Bouis, H. E. and R. M. Welch (2010). Biofortification-A Sustainable Agricultural Strategy for Reducing Micronutrient Malnutrition in the Global South. *Crop Sci* **50**(2): S20-S32.

Bovell-Benjamin, A. C., F. E. Viteri and L. H. Allen (1999). Sensory quality and lipid oxidaton of maize porridge as affected by iron amino acid chelate and EDTA. *J Food Science* **64**: 371.

Bovell-Benjamin, A. C., F. E. Viteri and L. H. Allen (2000). Iron absorption from ferrous bisglycinate and ferric trisglycinate in whole maize is regulated by iron status. *Am J Clin Nutr* **71**(6): 1563-1569.

Brandon, M. J., L. Y. Foo, L. J. Porter and P. Meredith (1982). Proanthocyanidins of Barley and Sorghum - Composition as a Function of Maturity of Barley Ears. *Phytochemistry* **21**(12): 2953-2957.

Briend, A. (2001). Highly nutrient-dense spreads: a new approach to delivering multiple micronutrients to high-risk groups. *Brit J Nutr* **85**: S175-S179.

Briend, A. and N. W. Solomons (2003). The evolving applications of spreads as a FOODlet for improving the diets of infants and young children. *Food Nutr Bull* **24**(3 Suppl): S34-38.

Brinch-Pedersen, H., F. Hatzack, E. Stoger, E. Arcalis, K. Pontopidan and P. B. Holm (2006). Heat-stable phytases in transgenic wheat (Triticum aestivum l.): Deposition pattern, thermostability, and phytate hydrolysis. *J Agr Food Chem* **54**(13): 4624-4632.

Brissot, P., T. L. Wright, W. L. Ma and R. A. Weisiger (1985). Efficient clearance of non-transferrin-bound iron by rat liver. Implications for hepatic iron loading in iron overload states. *J Clin Invest* **76**(4): 1463-1470.

Brooker, S., W. Akhwale, R. Pullan, B. Estambale, S. E. Clarke, R. W. Snow and P. J. Hotez (2007). Epidemiology of plasmodium-helminth co-infection in Africa: populations at risk, potential impact on anemia, and prospects for combining control. *Am J Trop Med Hyg* **77**(6 Suppl): 88-98.

Brune, M., L. Hallberg and A. B. Skanberg (1991). Determination of Iron-Binding Phenolic Groups in Foods. *J Food Sci* **56**(1): 128-&.

Brune, M., L. Rossander and L. Hallberg (1989a). Iron-Absorption - No Intestinal Adaptation to a High-Phytate Diet. *Am J Clin Nutr* **49**(3): 542-545.

Brune, M., L. Rossander and L. Hallberg (1989b). Iron absorption and phenolic compounds: importance of different phenolic structures. *Eur J Clin Nutr* **43**(8): 547-557.

Bruyeron, O., M. Denizeau, J. Berger and S. Treche (2010). Marketing complementary foods and supplements in Burkina Faso, Madagascar, and Vietnam: lessons learned from the Nutridev program. *Food Nutr Bull* **31**(2 Suppl): S154-167.

Burns, R. E. (1963) Methods of Tannin Analysis for Forage Crop Evaluation. In *Technical Bulletin no. 32*. Athens, GA: Georgia Agriculture Experiment Station.

Burns, R. E. (1971). Method for Estimation of Tannin in Grain Sorghum. *Agron J* **63**(3): 511-&.

Campion, B., F. Sparvoli, E. Doria, G. Tagliabue, I. Galasso, M. Fileppi, . . . E. Nielsen (2009). Isolation and characterisation of an lpa (low phytic acid) mutant in common bean (Phaseolus vulgaris L.). *Theor Appl Genet* **118**(6): 1211-1221.

Carpenter, C. E. and A. W. Mahoney (1992). Contributions of Heme and Nonheme Iron to Human-Nutrition. *Crit Rev Food Sci* **31**(4): 333-367.

Chang, S. Y., Z. W. Huang, Y. X. Ma, J. H. Piao, X. G. Yang, C. Zeder, . . . I. Egli (2012). Mixture of ferric sodium ethylenediaminetetraacetate (NaFeEDTA) and ferrous sulfate: An effective iron fortificant for complementary foods for young Chinese children. *Food Nutr Bull* **33**(2): 111-116.

Chen, J., X. Zhao, X. Zhang, S. Yin, J. Piao, J. Huo, . . . C. Chen (2005). Studies on the effectiveness of NaFeEDTA-fortified soy sauce in controlling iron deficiency: a population-based intervention trial. *Food Nutr Bull* **26**(2): 177-186; discussion 187-179.

Chen, Z., H. A. Meng, G. M. Xing, C. Y. Chen, Y. L. Zhao, G. A. Jia, . . . L. J. Wan (2006). Acute toxicological effects of copper nanoparticles in vivo. *Toxicol Lett* **163**(2): 109-120.

Chen, Z., T. E. Young, J. Ling, S. C. Chang and D. R. Gallie (2003). Increasing vitamin C content of plants through enhanced ascorbate recycling. *P Natl Acad Sci USA* **100**(6): 3525-3530.

Cheynier, V. (2005). Polyphenols in foods are more complex than often thought. *Am J Clin Nutr* **81**(1): 223s-229s.

Chiba, Y., J. H. Bryce, V. Goodfellow, J. MacKinlay, R. C. Agu, J. M. Brosnan, . . . B. Harrison (2012). Effect of germination temperatures on proteolysis of the gluten-free grains sorghum and millet during malting and mashing. *J Agric Food Chem* **60**(14): 3745-3753.

Chippaux, J. P., D. Schneider, A. Aplogan, J. L. Dyck and J. Berger (1991). [Effects od iron supplementation on malaria infection]. *Bull Soc Pathol Exot* **84**(1): 54-62.

Chirwa, R. M., V. D. Aggarwal, M. A. R. Phiri and A. R. E. Mwenda (2006). Experiences in implementing the bean seed strategy in Malawi. *J Sustain Agr* **29**(2): 43-69.

Choto, C. E., M. M. Morad and L. W. Rooney (1985). The Quality of Tortillas Containing Whole Sorghum and Pearled Sorghum Alone and Blended with Yellow Maize. *Cereal Chem* **62**(1): 51-55.

Chowdhury, S. and D. Punia (1997). Nutrient and antinutrient composition of pearl millet grains as affected by milling and baking. *Nahrung-Food* **41**(2): 105-107.

Christofides, A., K. P. Asante, C. Schauer, W. Sharieff, S. Owusu-Agyei and S. Zlotkin (2006). Multi-micronutrient Sprinkles including a low dose of iron provided as microencapsulated ferrous fumarate improves haematologic indices in anaemic children: a randomized clinical trial. *Matern Child Nutr* **2**(3): 169-180.

Christofides, A., C. Schauer and S. H. Zlotkin (2005). Iron deficiency anemia among children: Addressing a global public health problem within a Canadian context. *Paediatr Child Health* **10**(10): 597-601.

Clifford, M. N. (1999). Chlorogenic acids and other cinnamates - nature, occurrence and dietary burden. *J Sci Food Agric* **79**(3): 362-372.

Clifford, M. N. (2000). Anthocyanins - nature, occurrence and dietary burden. *J Sci Food Agric* **80**(7): 1063-1072.

Conner, A. J., T. R. Glare and J. P. Nap (2003). The release of genetically modified crops into the environment - Part II. Overview of ecological risk assessment. *Plant J* **33**(1): 19-46.

Consaul, J. R. and K. Lee (1983). Extrinsic Tagging in Iron Bioavailability Research - a Critical-Review. *J Agr Food Chem* **31**(4): 684-689.

Cook, J. D., S. A. Dassenko and S. R. Lynch (1991a). Assessment of the role of nonheme-iron availability in iron balance. *Am J Clin Nutr* **54**(4): 717-722.

Cook, J. D., S. A. Dassenko and P. Whittaker (1991b). Calcium supplementation: effect on iron absorption. *Am J Clin Nutr* **53**(1): 106-111.

Cook, J. D., C. A. Finch, R. Walker, Martinez.C, M. Layrisse and E. Monsen (1972). Food Iron-Absorption Measured by an Extrinsic Tag. *J Clin Invest* **51**(4): 805-&.

Cook, J. D., V. Minnich, C. V. Moore, Rasmusse.A, W. B. Bradley and C. A. Finch (1973). Absorption of Fortification Iron in Bread. *Am J Clin Nutr* **26**(8): 861-872.

Cook, J. D. and E. R. Monsen (1976a). Food iron absorption in human subjects. III. Comparison of the effect of animal proteins on nonheme iron absorption. *Am J Clin Nutr* **29**(8): 859-867.

Cook, J. D. and E. R. Monsen (1976b). Food iron absorption in man II. The effect of EDTA on absorption of dietary non-heme iron. *Am J Clin Nutr* **29**(6): 614-620.

Cook, J. D. and M. B. Reddy (2001). Effect of ascorbic acid intake on nonheme-iron absorption from a complete diet. *Am J Clin Nutr* **73**(1): 93-98.

Cook, J. D., M. B. Reddy and R. F. Hurrell (1995). The effect of red and white wines on nonheme-iron absorption in humans. *Am J Clin Nutr* **61**(4): 800-804.

Cook, J. D., S. S. Watson, K. M. Simpson, D. A. Lipschitz and B. S. Skikne (1984). The Effect of High Ascorbic-Acid Supplementation on Body Iron Stores. *Blood* **64**(3): 721-726.

Cosgrove, D. J. (1980) Inositolhexakisphosphates. In *Inositol phosphates*, pp. 26-43 [D. J. Cosgrove, editor]. Amsterdam: Elsevier Scientific Publishing Company.

Crawley, J. (2004). Reducing the burden of anemia in infants and young children in malaria-endemic countries of Africa: from evidence to action. *Am J Trop Med Hyg* **71**(2 Suppl): 25-34.

Crowley, C. R., N. W. Solomons and K. Schumann (2012). Targeted provision of oral iron: the evolution of a practical screening option. *Adv Nutr* **3**(4): 560-569.

Crutcher, J. M. and S. L. Hoffman (1996) Malaria. In *Medical Microbiology*, 4th ed., pp. 995-1008 [S. Baron, editor]. Galveston (TX): University of Texas Medical Branch.

Cureton, P. and A. Fasana (2009) The increasing incidence of celiac disease and the range of gluten-free products in the marketplace. In *Gluten-Free Food Science and Technology*, pp. 1-15 [E. Gallagher, editor]. Oxford: Blackwell Publishing Ltd.

David, S., L. Mukandala and J. Mafuru (2002). Seed availability, an ignored factor in crop varietal adoption studies: A case study of beans in Tanzania. *J Sustain Agr* **21**(2): 5-20.

Davidsson, L., A. Almgren, M. A. Juillerat and R. F. Hurrell (1995). Manganese absorption in humans: the effect of phytic acid and ascorbic acid in soy formula. *Am J Clin Nutr* **62**(5): 984-987.

Davidsson, L., T. Dimitriou, E. Boy, T. Walczyk and R. F. Hurrell (2002). Iron bioavailability from iron-fortified Guatemalan meals based on corn tortillas and black bean paste. *Am J Clin Nutr* **75**(3): 535-539.

Davidsson, L., T. Dimitriou, T. Walczyk and R. F. Hurrell (2001a). Iron absorption from experimental infant formulas based on pea (Pisum sativum)-protein isolate: the effect of phytic acid and ascorbic acid. *Br J Nutr* **85**(1): 59-63.

Davidsson, L., P. Galan, P. Kastenmayer, F. Cherouvrier, M. A. Juillerat, S. Hercberg and R. F. Hurrell (1994). Iron bioavailability studied in infants: the influence of phytic acid and ascorbic acid in infant formulas based on soy isolate. *Pediatr Res* **36**(6): 816-822.

Davidsson, L., P. Kastenmayer, H. Szajewska, R. F. Hurrell and D. Barclay (2000). Iron bioavailability in infants from an infant cereal fortified with ferric pyrophosphate or ferrous fumarate. *Am J Clin Nutr* **71**(6): 1597-1602.

Davidsson, L., S. A. Sarker, K. A. Jamil, S. Sultana and R. Hurrell (2009). Regular consumption of a complementary food fortified with ascorbic acid and ferrous fumarate or ferric pyrophosphate is as useful as ferrous sulfate in maintaining hemoglobin concentrations > 105 g/L in young Bangladeshi children. *Am J Clin Nutr* **89**(6): 1815-1820.

Davidsson, L., T. Walczyk, A. Morris and R. F. Hurrell (1998). Influence of ascorbic acid on iron absorption from an iron-fortified, chocolate-flavored milk drink in Jamaican children. *Am J Clin Nutr* **67**(5): 873-877.

Davidsson, L., T. Walczyk, N. Zavaleta and R. Hurrell (2001b). Improving iron absorption from a Peruvian school breakfast meal by adding ascorbic acid or Na2EDTA. *Am J Clin Nutr* **73**(2): 283-287.

De-Regil, L. M., P. S. Suchdev, G. E. Vist, S. Walleser and J. P. Pena-Rosas (2011). Home fortification of foods with multiple micronutrient powders for health and nutrition in children under two years of age. *Cochrane Database Syst Rev*(9): CD008959.

de Mast, Q., D. Syafruddin, S. Keijmel, T. O. Riekerink, O. Deky, P. B. Asih, . . . A. J. van der Ven (2010). Increased serum hepcidin and alterations in blood iron parameters associated with asymptomatic P. falciparum and P. vivax malaria. *Haematologica* **95**(7): 1068-1074.

de Mast, Q., E. C. van Dongen-Lases, D. W. Swinkels, A. E. Nieman, M. Roestenberg, P. Druilhe, . . . A. J. van der Ven (2009). Mild increases in serum hepcidin and interleukin-6 concentrations impair iron incorporation in haemoglobin during an experimental human malaria infection. *Brit J Haematol* **145**(5): 657-664.

de Oliveira, J. E., M. M. Scheid, I. D. Desai and S. Marchini (1996). Iron fortification of domestic drinking water to prevent anemia among low socioeconomic families in Brazil. *Int J Food Sci Nutr* **47**(3): 213-219.

de Paula, R. A. C. and M. Fisberg (2001). The use of sugar fortified with iron tris-glycinate chelate in the prevention of iron deficiency anemia in preschool children. *Arch Latinoam Nutr* **51**(1): 54-59.

de Pee, S. and M. W. Bloem (2009). Current and potential role of specially formulated foods and food supplements for preventing malnutrition among 6-to 23-month-old children and for treating moderate malnutrition among 6-to 59-month-old children. *Food Nutr Bull* **30**(3): 434-463.

de Pee, S., K. Kraemer, T. van den Briel, E. Boy, C. Grasset, R. Moench-Pfanner, . . . M. W. Bloem (2008). Quality criteria for micronutrient powder products: Report of a meeting organized by the World Food Programme and Sprinkles Global Health Initiative. *Food Nutr Bull* **29**(3): 232-241.

Demaneche, S., H. Sanguin, J. Pote, E. Navarro, D. Bernillon, P. Mavingui, . . . P. Simonet (2008). Antibiotic-resistant. soil bacteria in transigenic plant fields. *P Natl Acad Sci USA* **105**(10): 3957-3962.

Derman, D. P., D. Ballot, T. H. Bothwell, B. J. MacFarlane, R. D. Baynes, A. P. MacPhail, . . . F. Mayet (1987). Factors influencing the absorption of iron from soya-bean protein products. *Br J Nutr* **57**(3): 345-353.

Derman, D. P., T. H. Bothwell, J. D. Torrance, W. R. Bezwoda, A. P. Macphail, M. C. Kew, . . . R. W. Charlton (1980). Iron-Absorption from Maize (Zea-Mays) and Sorghum (Sorghum-Vulgare) Beer. *Brit J Nutr* **43**(2): 271-279.

Desai, M. R., J. V. Mei, S. K. Kariuki, K. A. Wannemuehler, P. A. Phillips-Howard, B. L. Nahlen, . . . F. O. ter Kuile (2003). Randomized, controlled trial of daily iron supplementation and intermittent sulfadoxine-pyrimethamine for the treatment of mild childhood anemia in western Kenya. *J Infect Dis* **187**(4): 658-666.

Dewey, K., J. Berger, J. S. Chen, C. M. Chen, S. de Pee, S. Huffman, . . . T. Y. S. R. V. M (2009a). Formulations for fortified complementary foods and supplements: Review of successful products for improving the nutritional status of infants and young children. *Food Nutr Bull* **30**(2): 239-255.

Dewey, K. G. and L. M. Baldiviez (2012). Safety of universal provision of iron through home fortification of complementary foods in malaria-endemic areas. *Adv Nutr* **3**(4): 555-559.

Dewey, K. G., Z. Y. Yang and E. Boy (2009b). Systematic review and meta-analysis of home fortification of complementary foods. *Matern Child Nutr* **5**(4): 283-321.

Dicko, M. H., H. Gruppen, A. S. Traore, W. J. H. Van Berkel and A. G. J. Voragen (2005). Evaluation of the effect of germination on phenolic compounds and antioxidant activities in sorghum varieties. *J Agr Food Chem* **53**(7): 2581-2588.

Dicko, M. H., R. Hilhorst, H. Gruppen, A. S. Traore, C. Laane, W. J. H. van Berkel and A. G. J. Voragen (2002). Comparison of content in phenolic compounds, polyphenol oxidase, and peroxidase in grains of fifty sorghum varieties from Burkina Faso. *J Agr Food Chem* **50**(13): 3780-3788.

Disler, P. B., S. R. Lynch, R. W. Charlton, J. D. Torrance, T. H. Bothwell, R. B. Walker and F. Mayet (1975). The effect of tea on iron absorption. *Gut* **16**(3): 193-200.

Dlamini, N. R., J. R. N. Taylor and L. W. Rooney (2007). The effect of sorghum type and processing on the antioxidant properties of African sorghum-based foods. *Food Chem* **105**(4): 1412-1419.

Doherty, C. P., S. E. Cox, A. J. Fulford, S. Austin, D. C. Hilmers, S. A. Abrams and A. M. Prentice (2008). Iron Incorporation and Post-Malaria Anaemia. *Plos One* **3**(5): e2133.

Domellof, M., R. J. Cohen, K. G. Dewey, O. Hernell, L. L. Rivera and B. Lonnerdal (2001). Iron supplementation of breast-fed Honduran and Swedish infants from 4 to 9 months of age. *J Pediatr* **138**(5): 679-687.

Dorsch, J. A., A. Cook, K. A. Young, J. M. Anderson, A. T. Bauman, C. J. Volkmann, . . . V. Raboy (2003). Seed phosphorus and inositol phosphate phenotype of barley low phytic acid genotypes. *Phytochemistry* **62**(5): 691-706.

Drakakaki, G., S. Marcel, R. P. Glahn, E. K. Lund, S. Pariagh, R. Fischer, . . . E. Stoger (2005). Endosperm-specific co-expression of recombinant soybean ferritin and Aspergillus phytase in maize results in significant increases in the levels of bioavailable iron. *Plant Mol Biol* **59**(6): 869-880.

Dresow, B., D. Petersen, R. Fischer and P. Nielsen (2008). Non-transferrin-bound iron in plasma following administration of oral iron drugs. *BioMetals* **21**(3): 273-276.

Drynan, J. W., M. N. Clifford, J. Obuchowicz and N. Kuhnert (2010). The chemistry of low molecular weight black tea polyphenols. *Nat Prod Rep* **27**(3): 417-462.

Dykes, L., G. C. Peterson, W. L. Rooney and L. W. Rooney (2011). Flavonoid composition of lemon-yellow sorghum genotypes. *Food Chem* **128**(1): 173-179.

Dykes, L. and L. W. Rooney (2006). Sorghum and millet phenols and antioxidants. *J Cereal Sci* **44**(3): 236-251.

Dykes, L. and L. W. Rooney (2007). Phenolic compounds in cereal grains and their health benefits. *Cereal Food World* **52**(3): 105-111.

Dykes, L., L. W. Rooney, R. D. Waniska and W. L. Rooney (2005). Phenolic compounds and antioxidant activity of sorghum grains of varying genotypes. *J Agric Food Chem* **53**(17): 6813-6818.

Dykes, L., L. M. Seitz, W. L. Rooney and L. W. Rooney (2009). Flavonoid composition of red sorghum genotypes. *Food Chem* **116**(1): 313-317.

Earp, C. F., J. O. Akingbala, S. H. Ring and L. W. Rooney (1981). Evaluation of Several Methods to Determine Tannins in Sorghums with Varying Kernel Characteristics. *Cereal Chem* **58**(3): 234-238.

Egli, I., L. Davidsson, M. A. Juillerat, D. Barclay and R. Hurrell (2003). Phytic acid degradation in complementary foods using phytase naturally occurring in whole grain cereals. *J Food Sci* **68**(5): 1855-1859.

Eke, R. A., L. N. Chigbu and N. W. (2006). High Prevalence of Asymptomatic Plasmodium Infection in a Suburb of Aba Town, Nigeria. *Ann Afr Med* **5**(1): 42-45.

El Hag, M. E., A. H. El Tinay and N. E. Yousif (2002). Effect of fermentation and dehulling on starch, total polyphenols, phytic acid content and in vitro protein digestibility of pearl millet. *Food Chem* **77**(2): 193-196.

Elyas, S. H. A., A. H. El Tinay, N. E. Yousif and E. A. E. Elshelkh (2002). Effect of natural fermentation on nutritive value and in vitro protein digestibility of pearl millet. *Food Chem* **78**(1): 75-79.

Escarpa, A. and M. C. Gonzalez (2001). An overview of analytical chemistry of phenolic compounds in foods. *Crit Rev Anal Chem* **31**(2): 57-139.

European Food Safety Authority (2010) Scientific opinion on the use of ferric sodium EDTA as a source of iron added for nutritional purposes to foods for the general population (including food supplements) and to food for particular nutritional uses. Parma: European Food Safety Authority.

European Union (2010) Commission Decision of 14 June 2010: authorising the placing on the market of Ferric Sodium EDTA as a novel food ingredient under Regulation (EC) No 258/97 of the European Parliament and of the Council. http://eur-lex.europa.eu/LexUriServ/LexUriServ.do?uri=OJ:L:2010:149:0016:0019:EN:PDF (accessed January 2013).

Eyzaguirre, R. Z., K. Nienaltowska, L. E. Q. de Jong, B. B. E. Hasenack and M. J. R. Nout (2006). Effect of food processing of pearl millet (Pennisetum glaucum) IKMP-5 on the level of phenolics, phytate, iron and zinc. *J Sci Food Agric* **86**(9): 1391-1398.

Faber, M., J. D. Kvalsvig, C. J. Lombard and A. J. S. Benade (2005). Effect of a fortified maize-meal porridge on anemia, micronutrient status, and motor development of infants. *Am J Clin Nutr* **82**(5): 1032-1039.

Fairweather-Tait, S., S. Lynch, C. Hotz, R. Hurrell, L. Abrahamse, S. Beebe, . . . T. P. Trinidad (2005). The usefulness of in vitro models to predict the bioavailability of iron and zinc: A consensus statement from the HarvestPlus expert consultation. *Int J Vitam Nutr Res* **75**(6): 371-374.

Fairweather-Tait, S., I. Phillips, G. Wortley, L. Harvey and R. Glahn (2007). The use of solubility, dialyzability, and Caco-2 cell methods to predict iron bioavailability. *Int J Vitam Nutr Res* **77**(3): 158-165.

Fairweather-Tait, S. J., G. M. Wortley, B. Teucher and J. Dainty (2001). Iron absorption from a breakfast cereal: effects of EDTA compounds and ascorbic acid. *Int J Vitam Nutr Res* **71**(2): 117-122.

FAO (1995) *Sorghum and millets in human nutrition.* Rome; Lanham, MD: Food and Agriculture Organization of the United Nations; UNIPUB distributor.

FAO (2009) FAOSTAT: Food Security Statistics. http://faostat.fao.org/ (accessed October 2012).

Favier, J.-C. (1989) Valeur nutritive des céréales au cours de leurs transformations. In *Céréales au régions chaudes*, pp. 285-297. Paris: John Libbey Eurotext.

Ferguson, E. L., R. S. Gibson, L. U. Thompson and S. Ounpuu (1989). Dietary Calcium, Phytate, and Zinc Intakes and the Calcium, Phytate, and Zinc Molar Ratios of the Diets of a Selected Group of East-African Children. *Am J Clin Nutr* **50**(6): 1450-1456.

Fetene, M., Okori P., Gudu S., Mneney E. and T. K. (2011) Delivering New Sorghum and Finger Millet Innovations for Food Security and Improving Livelihoods in Eastern Africa. http://mahider.ilri.org/bitstream/handle/10568/10811/Project1_Sorghum.pdf?sequence=6 (accessed January 2013).

Fidler, M. C., L. Davidsson, T. Walczyk and R. F. Hurrell (2003a). Iron absorption from fish sauce and soy sauce fortified with sodium iron EDTA. *Am J Clin Nutr* **78**(2): 274-278.

Fidler, M. C., L. Davidsson, C. Zeder, T. Walczyk and R. F. Hurrell (2003b). Iron absorption from ferrous fumarate in adult women is influenced by ascorbic acid but not by Na(2)EDTA. *Brit J Nutr* **90**(6): 1081-1085.

Fidler, M. C., L. Davidsson, C. Zeder, T. Walczyk, I. Marti and R. F. Hurrell (2004a). Effect of ascorbic acid and particle size on iron absorption from ferric pyrophosphate in adult women. *Int J Vitam Nutr Res* **74**(4): 294-300.

Fidler, M. C., A. Krzystek, T. Walczyk and R. F. Hurrell (2004b). Photostability of sodium iron ethylenediaminetetraacetic acid (NaFeEDTA) in stored fish sauce and soy sauce. *J Food Sci* **69**(9): S380-S383.

Fidler, M. C., T. Walczyk, L. Davidsson, C. Zeder, N. Sakaguchi, L. R. Juneja and R. F. Hurrell (2004c). A micronised, dispersible ferric pyrophosphate with high relative bioavailability in man. *Brit J Nutr* **91**(1): 107-112.

Figueroacolon, R., J. H. Elwell, E. Jackson, J. W. Osborne and S. J. Fomon (1989). Failure to Label Baboon Milk Intrinsically with Iron. *J Pediatr Gastr Nutr* **9**(4): 521-523.

Fisberg, M., J. A. Pellegrin Braga, P. E. Kliamca, A. M. Amaral Ferreira and M. Berezowski (1995). Utilization of Petit Suisse cheese in the prevention of nutritional anemia in preschool children. *Clin Pediat (Brazil)* **19**: 14.

Fleming, D. J., P. F. Jacques, G. E. Dallal, K. L. Tucker, P. W. Wilson and R. J. Wood (1998). Dietary determinants of iron stores in a free-living elderly population: The Framingham Heart Study. *Am J Clin Nutr* **67**(4): 722-733.

Forbes, A. L., C. E. Adams, M. J. Arnaud, C. O. Chichester, J. D. Cook, B. N. Harrison, . . . P. Whittaker (1989). Comparison of Invitro, Animal, and Clinical Determinations of Iron Bioavailability - International-Nutritional-Anemia Consultative-Group-Task-Force Report on Iron Bioavailability. *Am J Clin Nutr* **49**(2): 225-238.

Fox, T. E., J. Eagles and S. J. Fairweather-Tait (1998). Bioavailability of iron glycine as a fortificant in infant foods. *Am J Clin Nutr* **67**(4): 664-668.

Friedman, M. (1997). Chemistry, biochemistry, and dietary role of potato polyphenols. A review. *J Agr Food Chem* **45**(5): 1523-1540.

Fulda, S. (2010). Resveratrol and derivatives for the prevention and treatment of cancer. *Drug Discov Today* **15**(17-18): 757-765.

Gaitan, D., S. Flores, P. Saavedra, C. Miranda, M. Olivares, M. Arredondo, . . . F. Pizarro (2011). Calcium does not inhibit the absorption of 5 milligrams of nonheme or heme iron at doses less than 800 milligrams in nonpregnant women. *J Nutr* **141**(9): 1652-1656.

Galiba, M., L. W. Rooney, R. D. Waniska and F. R. Miller (1987). The Preparation of Sorghum and Millet Couscous in West-Africa. *Cereal Food World* **32**(12): 878-&.

Ganz, T. (2007). Molecular control of iron transport. *J Am Soc Nephrol* **18**(2): 394-400.

Gartlehner, G., R. A. Hansen, D. Nissman, K. N. Lohr and T. S. Carey (2006) *Criteria for Distinguishing Effectiveness From Efficacy Trials in Systematic Reviews*. Rockville (MD): Agency for Healthcare Research and Quality (US).

Gera, T., H. S. Sachdev and E. Boy (2012). Effect of iron-fortified foods on hematologic and biological outcomes: systematic review of randomized controlled trials. *Am J Clin Nutr* **96**(2): 309-324.

Ghosh, K. (2007). Pathogenesis of anemia in malaria: a concise review. *Parasitol Res* **101**(6): 1463-1469.

Gibson, R. S. (1997). Technological approaches to combatting iron deficiency. *Eur J Clin Nutr* **51 Suppl 4**: S25-27.

Gibson, R. S., K. B. Bailey, M. Gibbs and E. L. Ferguson (2010). A review of phytate, iron, zinc, and calcium concentrations in plant-based complementary foods used in low-income countries and implications for bioavailability. *Food Nutr Bull* **31**(2 Suppl): S134-146.

Gibson, R. S., E. L. Ferguson and J. Lehrfeld (1998). Complementary foods for infant feeding in developing countries: their nutrient adequacy and improvement. *Eur J Clin Nutr* **52**(10): 764-770.

Gibson, R. S. and C. Hotz (2001). Dietary diversification/modification strategies to enhance micronutrient content and bioavailability of diets in developing countries. *Br J Nutr* **85 Suppl 2**: S159-166.

Gibson, S. A. (1999). Iron intake and iron status of preschool children: associations with breakfast cereals, vitamin C and meat. *Public Health Nutr* **2**(4): 521-528.

Gillooly, M., T. H. Bothwell, R. W. Charlton, J. D. Torrance, W. R. Bezwoda, A. P. Macphail, . . . F. Mayet (1984). Factors Affecting the Absorption of Iron from Cereals. *Brit J Nutr* **51**(1): 37-46.

Gillooly, M., T. H. Bothwell, J. D. Torrance, A. P. MacPhail, D. P. Derman, W. R. Bezwoda, . . . F. Mayet (1983). The effects of organic acids, phytates and polyphenols on the absorption of iron from vegetables. *Br J Nutr* **49**(3): 331-342.

Giorgini, E., M. Fisberg, R. A. C. De Paula, A. M. A. Ferreira, J. Valle and J. A. P. Braga (2001). The use of sweet rolls fortified with iron bis-glycinate chelate in the prevention of iron deficiency anemia in preschool children. *Arch Latinoam Nutr* **51**(1): 48-53.

Giovannini, M., D. Sala, M. Usuelli, L. Livio, G. Francescato, M. Braga, . . . E. Riva (2006). Double-blind, placebo-controlled trial comparing effects of supplementation with two different combinations of micronutrients delivered as sprinkles on growth, anemia, and iron deficiency in cambodian infants. *J Pediatr Gastroenterol Nutr* **42**(3): 306-312.

Githeko, A. K., A. D. Brandling-Bennett, M. Beier, C. M. Mbogo, F. K. Atieli, M. L. Owaga, . . . F. H. Collins (1993). Confirmation that Plasmodium falciparum has aperiodic infectivity to Anopheles gambiae. *Med. Vet. Entomol.* **7**(4): 373-376.

Goma, J., L. Renia, F. Miltgen and D. Mazier (1996). Iron overload increases hepatic development of Plasmodium yoelii in mice. *Parasitology* **112 (Pt 2)**: 165-168.

Gordeuk, V. R., P. E. Thuma, G. M. Brittenham, S. Zulu, G. Simwanza, A. Mhangu, . . . D. Parry (1992). Iron chelation with desferrioxamine B in adults with asymptomatic Plasmodium falciparum parasitemia. *Blood* **79**(2): 308-312.

Graham, R. D., R. M. Welch and H. E. Bouis (2001). Addressing micronutrient malnutrition through enhancing the nutritional quality of staple foods: Principles, perspectives and knowledge gaps. *Adv Agron* **70**: 77-142.

Graham, S. M. (2010). Nontyphoidal salmonellosis in Africa. *Curr Opin Infect Dis* **23**(5): 409-414.

Greiner, R. and U. Konietzny (1999). Improving enzymatic reduction of myo-inositol phosphates with inhibitory effects on mineral absorption in black beans (Phaseolus vulgaris var. Preto). *J Food Process Pres* **23**(3): 249-261.

Grusak, M. A. (1997). Intrinsic stable isotope labeling of plants for nutritional investigations in humans. *J Nutr Biochem* **8**(4): 164-171.

Grusak, M. A. and I. Cakmak (2005) Methods to improve the crop-delivery of minerals to humans and livestock. In *Plant Nutritional Genomics*, pp. 265-286 [M. R. Broadley and P. J. White, editors]. Oxford: Blackwell.

Gu, L. W., M. Kelm, J. F. Hammerstone, G. Beecher, D. Cunningham, S. Vannozzi and R. L. Prior (2002). Fractionation of polymeric procyanidins from lowbush blueberry and quantification of procyanidins in selected foods with an optimized normal-phase HPLC-MS fluorescent detection method. *J Agr Food Chem* **50**(17): 4852-4860.

Gu, L. W., M. A. Kelm, J. F. Hammerstone, G. Beecher, J. Holden, D. Haytowitz and R. L. Prior (2003). Screening of foods containing proanthocyanidins and their structural characterization using LC-MS/MS and thiolytic degradation. *J Agr Food Chem* **51**(25): 7513-7521.

Gujer, R., D. Magnolato and R. Self (1986). Glucosylated Flavonoids and Other Phenolic-Compounds from Sorghum. *Phytochemistry* **25**(6): 1431-1436.

Gupta, R. K. and E. Haslam (1978). Plant Proanthocyanidins .5. Sorghum Polyphenols. *J Chem Soc Perk T 1*(8): 892-896.

Gupta, S. K., G. Velu, K. N. Rai and K. Sumalini (2009). Association of grain iron and zinc content with grain yield and other traits in pearl millet (*pennisetum glaucum* (L.) R.BR.). *Crop Improvment* **36**(2): 4-7.

Guttieri, M., D. Bowen, J. A. Dorsch, V. Raboy and E. Souza (2003). Identification and characterization of a low phytic acid wheat. *Crop Sci* **44**(2): 418-424.

Guttieri, M. J., K. M. Peterson and E. J. Souza (2006). Agronomic performance of low phytic acid wheat. *Crop Sci* **46**(6): 2623-2629.

Gwamaka, M., J. D. Kurtis, B. E. Sorensen, S. Holte, R. Morrison, T. K. Mutabingwa, . . . P. E. Duffy (2012). Iron deficiency protects against severe Plasmodium falciparum malaria and death in young children. *Clin Infect Dis* **54**(8): 1137-1144.

Haas, J. D., J. L. Beard, L. E. Murray-Kolb, A. M. del Mundo, A. Felix and G. B. Gregorio (2005). Iron-biofortified rice improves the iron stores of nonanemic Filipino women. *J Nutr* **135**(12): 2823-2830.

Hahn, D. H., J. M. Faubion and L. W. Rooney (1983). Sorghum Phenolic-Acids, Their High-Performance Liquid-Chromatography Separation and Their Relation to Fungal Resistance. *Cereal Chem* **60**(4): 255-259.

Hallberg, K. and L. Hulthen (2000). Prediction of dietary iron adsorption: an algorithm for calculating absorption and bioavailability of dietary iron (vol 71, pg 1147, 2000). *Am J Clin Nutr* **72**(5): 1242-1242.

Hallberg, L. (1981). Bioavailability of Dietary Iron in Man. *Annu Rev Nutr* **1**: 123-147.

Hallberg, L. and E. Björn-Rasmussen (1981). Measurement of iron absorption from meals contaminated with iron. *Am J Clin Nutr* **34**(12): 2808-2815.

Hallberg, L., E. Björn-Rasmussen, L. Rossander, R. Suwanik, R. Pleehachinda and M. Tuntawiroon (1983). Iron absorption from some Asian meals containing contamination iron. *Am J Clin Nutr* **37**(2): 272-277.

Hallberg, L., M. Brune, M. Erlandsson, A. S. Sandberg and L. Rossander-Hulten (1991). Calcium: effect of different amounts on nonheme- and heme-iron absorption in humans. *Am J Clin Nutr* **53**(1): 112-119.

Hallberg, L., M. Brune and L. Rossander (1986). Effect of ascorbic acid on iron absorption from different types of meals. Studies with ascorbic-acid-rich foods and synthetic ascorbic acid given in different amounts with different meals. *Hum Nutr Appl Nutr* **40**(2): 97-113.

Hallberg, L., M. Brune and L. Rossander (1989a). Iron-Absorption in Man - Ascorbic-Acid and Dose-Dependent Inhibition by Phytate. *Am J Clin Nutr* **49**(1): 140-144.

Hallberg, L. and L. Rossander (1982). Effect of different drinks on the absorption of non-heme iron from composite meals. *Hum Nutr Appl Nutr* **36**(2): 116-123.

Hallberg, L. and L. Rossander (1984). Improvement of iron nutrition in developing countries: comparison of adding meat, soy protein, ascorbic acid, citric acid, and ferrous sulphate on iron absorption from a simple Latin American-type of meal. *Am J Clin Nutr* **39**(4): 577-583.

Hallberg, L., L. Rossanderhulthen and E. Gramatkovski (1989b). Iron Fortification of Flour with a Complex Ferric Ortho-Phosphate. *Am J Clin Nutr* **50**(1): 129-135.

Hallfors, D., H. Cho, V. Sanchez, S. Khatapoush, H. M. Kim and D. Bauer (2006). Efficacy vs effectiveness trial results of an indicated "model" substance abuse program: implications for public health. *Am J Public Health* **96**(12): 2254-2259.

Hama, F., C. Icard-Verniere, J. P. Guyot, C. Picq, B. Diawara and C. Mouquet-Rivier (2011). Changes in micro- and macronutrient composition of pearl millet and white sorghum during in field versus laboratory decortication. *J Cereal Sci* **54**(3): 425-433.

Hama, F., C. Icard-Verniere, J. P. Guyot, I. Rochette, B. Diawara and C. Mouquet-Rivier (2012). Potential of non-GMO biofortified pearl millet (Pennisetum glaucum) for increasing iron and zinc content and their estimated bioavailability during abrasive decortication. *Int J Food Sci Technol* **47**(8): 1660-1668.

Hansen, M., S. B. Baech, A. D. Thomsen, I. Tetens and B. Sandstrom (2005). Long-term intake of iron fortified wholemeal rye bread appears to benefit iron status of young women. *J Cereal Sci* **42**(2): 165-171.

Harlan, J. R. and J. M. J. Dewet (1972). Simplified Classification of Cultivated Sorghum. *Crop Sci* **12**(2): 172-&.

Harland, B. F., O. L. Oke and R. Felix-Philips (1988). Preliminary studies on the phytate content of Nigerian foods. *J Food Compos Anal* **1**: 202-205.

Harrington, M., C. Hotz, C. Zeder, G. O. Polvo, S. Villalpando, M. B. Zimmermann, . . . R. F. Hurrell (2011). A comparison of the bioavailability of ferrous fumarate and ferrous sulfate in non-anemic Mexican women and children consuming a sweetened maize and milk drink. *Eur J Clin Nutr* **65**(1): 20-25.

HarvestPlus (2009a) Iron pearl millet. http://www.unscn.org/layout/modules/resources/files/HarvestPlus_Pearl_Millet_Strateg y.pdf (accessed October 2012).

HarvestPlus (2009b) Product development & delivery design and implementation. http://www.harvestplus.org/sites/default/files/PDD%20Design%20and%20Implementation.pdf (accessed November 2012).

HarvestPlus (2009c) Project Portofolio: High Iron Pearl Millet for India. http://cgmap.cgiar.org/docsRepository/documents/MTPProjects/2009-2011/HPCP_2009-2011_06.PDF (accessed October 2012).

HarvestPlus (2011a) HarvestPlus Impact Pathway. http://www.harvestplus.org/content/harvestplus-impact-pathway (accessed November 2012).

HarvestPlus (2011b) Pearl millet set for release in 2012. http://www.harvestplus.org/content/pearl-millet-set-release-2012 (accessed October 2012).

Harvey, P. W., P. F. Heywood, M. C. Nesheim, K. Galme, M. Zegans, J. P. Habicht, . . . et al. (1989). The effect of iron therapy on malarial infection in Papua New Guinean schoolchildren. *Am J Trop Med Hyg* **40**(1): 12-18.

Harvey, P. W. J., P. B. Dexter and I. Darnton-Hill (2000). The impact of consuming iron from non-food sources on iron status in developing countries. *Public Health Nutr* **3(4)**: 375-383.

Haslam, E. (2007). Vegetable tannins - Lessons of a phytochemical lifetime. *Phytochemistry* **68**(22-24): 2713-2721.

Hemanalini, G., K. P. Umapathy, J. R. Rao and G. Saraswathi (1980). Nutritional-Evaluation of Sprouted Ragi. *Nutr Rep Int* **22**(2): 271-277.

Hentze, M. W., M. U. Muckenthaler and N. C. Andrews (2004). Balancing acts: molecular control of mammalian iron metabolism. *Cell* **117**(3): 285-297.

Herrmann, K. M. and L. M. Weaver (1999). The shikimate pathway. *Annu Rev Plant Phys* **50**: 473-503.

Hershko, C. (2007). Mechanism of iron toxicity. *Food Nutr Bull* **28**(4 Suppl): S500-509.

Hershko, C. and T. E. Peto (1988). Deferoxamine inhibition of malaria is independent of host iron status. *J Exp Med* **168**(1): 375-387.

Hertrampf, E. and M. Olivares (2004). Iron amino acid chelates. *Int J Vitam Nutr Res* **74**(6): 435-443.

Hider, R. C., Z. D. Liu and H. H. Khodr (2001). Metal chelation of polyphenols. *Methods Enzymol* **335**: 190-203.

Hilty, F. M., M. Arnold, M. Hilbe, A. Teleki, J. T. N. Knijnenburg, F. Ehrensperger, . . . M. B. Zimmermann (2010). Iron from nanocompounds containing iron and zinc is highly bioavailable in rats without tissue accumulation. *Nat Nanotechnol* **5**(5): 374-380.

Hilty, F. M., A. Teleki, F. Krumeich, R. Buchel, R. F. Hurrell, S. E. Pratsinis and M. B. Zimmermann (2009). Development and optimization of iron- and zinc-containing nanostructured powders for nutritional applications. *Nanotechnology* **20**(47).

Hirve, S., S. Bhave, A. Bavdekar, S. Naik, A. Pandit, C. Schauer, . . . S. Zlotkin (2007). Low dose 'Sprinkles'-- an innovative approach to treat iron deficiency anemia in infants and young children. *Indian Pediatr* **44**(2): 91-100.

Hoppe, C., G. S. Andersen, S. Jacobsen, C. Molgaard, H. Friis, P. T. Sangild and K. F. Michaelsen (2008). The use of whey or skimmed milk powder in fortified blended foods for vulnerable groups. *J Nutr* **138**(1): 145S-161S.

Hoppe, C., C. Molgaard and K. F. Michaelsen (2006). Cow's milk and linear growth in industrialized and developing countries. *Annu Rev Nutr* **26**: 131-173.

Horrocks, P., R. A. Pinches, S. J. Chakravorty, J. Papakrivos, Z. Christodoulou, S. A. Kyes, . . . C. I. Newbold (2005). PfEMP1 expression is reduced on the surface of knobless Plasmodium falciparum infected erythrocytes. *J Cell Sci* **118**(Pt 11): 2507-2518.

Hotz, C. and B. McClafferty (2007). From harvest to health: challenges for developing biofortified staple foods and determining their impact on micronutrient status. *Food Nutr Bull* **28**(2 Suppl): S271-279.

Hotz, C., M. Porcayo, G. Onofre, A. Garcia-Guerra, T. Elliott, S. Jankowski and T. Greiner (2008). Efficacy of iron-fortified Ultra Rice in improving the iron status of women in Mexico. *Food Nutr Bull* **29**(2): 140-149.

Howard, C. T., U. S. McKakpo, I. A. Quakyi, K. M. Bosompem, E. A. Addison, K. Sun, . . . R. D. Semba (2007). Relationship of hepcidin with parasitemia and anemia among patients with uncomplicated Plasmodium falciparum malaria in Ghana. *Am J Trop Med Hyg* **77**(4): 623-626.

Huang, J., J. Sun, W. X. Li, L. J. Wang, A. X. Wang, J. S. Huo, . . . C. M. Chen (2009). Efficacy of different iron fortificants in wheat flour in controlling iron deficiency. *Biomedical and environmental sciences : BES* **22**(2): 118-121.

Hulse, J. H., E. M. Laing and O. E. Pearson (1980) *Sorghum and the millets : their composition and nutritive value*. London ; New York: Academic Press.

Hulten, L., E. Gramatkovski, A. Gleerup and L. Hallberg (1995). Iron absorption from the whole diet. Relation to meal composition, iron requirements and iron stores. *Eur J Clin Nutr* **49**(11): 794-808.

Hunt, J. R. (2003). High-, but not low-bioavailability diets enable substantial control of women's iron absorption in relation to body iron stores, with minimal adaptation within several weeks. *Am J Clin Nutr* **78**(6): 1168-1177.

Hunt, J. R., S. K. Gallagher and L. K. Johnson (1994). Effect of Ascorbic-Acid on Apparent Iron-Absorption by Women with Low Iron Stores. *Am J Clin Nutr* **59**(6): 1381-1385.

Hunt, J. R., L. M. Mullen, G. I. Lykken, S. K. Gallagher and F. H. Nielsen (1990). Ascorbic acid: effect on ongoing iron absorption and status in iron-depleted young women. *Am J Clin Nutr* **51**(4): 649-655.

Hunt, J. R. and Z. K. Roughead (2000). Adaptation of iron absorption in men consuming diets with high or low iron bioavailability. *Am J Clin Nutr* **71**(1): 94-102.

Huo, J., J. Sun, H. Miao, B. Yu, T. Yang, Z. Liu, . . . Y. Li (2002). Therapeutic effects of NaFeEDTA-fortified soy sauce in anaemic children in China. *Asia Pac J Clin Nutr* **11**(2): 123-127.

Hurrell, R. (1984) Bioavailability of different iron compounds to fortify formulas and cereals: technological problems. In *Iron nutrition in infancy and childhood*, pp. 39-53 [A. Steckel, editor]. New York: Raven Press.

Hurrell, R. (1985) Types of iron fortificants. Nonelemntal sources. In *Iron fortification of foods*, pp. 39-53 [F. M. Clydesdale and K. L. Wiemer, editors]. Orlando: Academic Press.

Hurrell, R. (1999) Iron. In *The mineral fortification of foods*, pp. 54-93 [R. Hurrell, editor]. Surrey, UK: Leatherhead Food RA.

Hurrell, R. (2002a). How to ensure adequate iron absorption from iron-fortified food. *Nutr Rev* **60**(7): S7-S15.

Hurrell, R. (2007). Linking the bioavailability of iron compounds to the efficacy of iron-fortified foods. *Int J Vitam Nutr Res* **77**(3): 166-173.

Hurrell, R. (2010a). Iron and malaria: absorption, efficacy and safety. *Int J Vitam Nutr Res* **80**(4-5): 279-292.

Hurrell, R. (2010b). Use of ferrous fumarate to fortify foods for infants and young children. *Nutr Rev* **68**(9): 522-530.

Hurrell, R., T. Bothwell, J. D. Cook, O. Dary, L. Davidsson, S. Fairweather-Tait, . . . P. Whittaker (2002). The usefulness of elemental iron for cereal flour fortification: A SUSTAIN task force report. *Nutr Rev* **60**(12): 391-406.

Hurrell, R. and I. Egli (2010). Iron bioavailability and dietary reference values. *Am J Clin Nutr* **91**(5): 1461S-1467S.

Hurrell, R., P. Ranum, S. de Pee, R. Biebinger, L. Hulthen, Q. Johnson and S. Lynch (2010). Revised recommendations for iron fortification of wheat flour and an evaluation of the expected impact of current national wheat flour fortification programs. *Food Nutr Bull* **31**(1 Suppl): S7-21.

Hurrell, R. F. (1997). Preventing iron deficiency through food fortification. *Nutr Rev* **55**(6): 210-222.

Hurrell, R. F. (2002b). Fortification: Overcoming technical and practical barriers. *J Nutr* **132**(4): 806-812.

Hurrell, R. F. (2004a). Improved technologies enhance the efficacy of food iron fortification. *Int J Vitam Nutr Res* **74**(6): 385.

Hurrell, R. F. (2004b). Phytic acid degradation as a means of improving iron absorption. *Int J Vitam Nutr Res* **74**(6): 445-452.

Hurrell, R. F. (2011). Safety and efficacy of iron supplements in malaria-endemic areas. *Ann Nutr Metab* **59**(1): 64-66.

Hurrell, R. F., D. E. Furniss, J. Burri, P. Whittaker, S. R. Lynch and J. D. Cook (1989a). Iron fortification of infant cereals: a proposal for the use of ferrous fumarate or ferrous succinate. *Am J Clin Nutr* **49**(6): 1274-1282.

Hurrell, R. F., M. A. Juillerat, M. B. Reddy, S. R. Lynch, S. A. Dassenko and J. D. Cook (1992). Soy protein, phytate, and iron absorption in humans. *Am J Clin Nutr* **56**(3): 573-578.

Hurrell, R. F., S. R. Lynch, T. P. Trinidad, S. A. Dassenko and J. D. Cook (1988). Iron absorption in humans: bovine serum albumin compared with beef muscle and egg white. *Am J Clin Nutr* **47**(1): 102-107.

Hurrell, R. F., S. R. Lynch, T. P. Trinidad, S. A. Dassenko and J. D. Cook (1989b). Iron absorption in humans as influenced by bovine milk proteins. *Am J Clin Nutr* **49**(3): 546-552.

Hurrell, R. F., M. Reddy and J. D. Cook (1999). Inhibition of non-haem iron absorption in man by polyphenolic-containing beverages. *Brit J Nutr* **81**(4): 289-295.

Hurrell, R. F., M. B. Reddy, J. Burri and J. D. Cook (2000a). An evaluation of EDTA compounds for iron fortification of cereal-based foods. *Br J Nutr* **84**(6): 903-910.

Hurrell, R. F., M. B. Reddy, J. Burri and J. D. Cook (2000b). An evaluation of EDTA compounds for iron fortification of cereal-based foods. *Brit J Nutr* **84**(6): 903-910.

Hurrell, R. F., M. B. Reddy, S. A. Dassenko, J. D. Cook and D. Shepherd (1991). Ferrous Fumarate Fortification of a Chocolate Drink Powder. *Brit J Nutr* **65**(2): 271-283.

Hurrell, R. F., M. B. Reddy, M. A. Juillerat and J. D. Cook (2003). Degradation of phytic acid in cereal porridges improves iron absorption by human subjects. *Am J Clin Nutr* **77**(5): 1213-1219.

Hurrell, R. F., S. Ribas and L. Davidsson (1994). NaFe3+EDTA as a food fortificant: influence on zinc, calcium and copper metabolism in the rat. *Br J Nutr* **71**(1): 85-93.

Hutchinson, C., W. Al-Ashgar, D. Y. Liu, R. C. Hider, J. J. Powell and C. A. Geissler (2004). Oral ferrous sulphate leads to a marked increase in pro-oxidant nontransferrin-bound iron. *Eur J Clin Invest* **34**(11): 782-784.

ICRISAT/FAO (1996) *The world sorghum and millet economics: facts, trends and outlook.* Patancheru, India, Rome: ICRISAT/FAO.

International Atomic Energy Agency (2012) *Assessment of iron bioavailability in humans using stable isotope techniques.* Vienna: International Atomic Energy Agency.

Iost, C., J. J. Name, R. B. Jeppsen and E. D. Ashmead (1998). Repleting hemoglobin in iron deficiency anemia in young children through liquid milk fortification with bioavailable iron amino acid chelate. *J Am Coll Nutr* **17**(2): 187-194.

Ip, H., S. M. Z. Hyder, F. Haseen, M. Rahman and S. H. Zlotkin (2009). Improved adherence and anaemia cure rates with flexible administration of micronutrient Sprinkles: a new public health approach to anaemia control. *Eur J Clin Nutr* **63**(2): 165-172.

Jack, S. J., K. Ou, M. Chea, L. Chhin, R. Devenish, M. Dunbar, . . . R. S. Gibson (2012). Effect of micronutrient sprinkles on reducing anemia: a cluster-randomized effectiveness trial. *Arch. Pediatr. Adolesc. Med.* **166**(9): 842-850.

Jani, P., G. W. Halbert, J. Langridge and A. T. Florence (1990). Nanoparticle Uptake by the Rat Gastrointestinal Mucosa - Quantitation and Particle-Size Dependency. *J Pharm Pharmacol* **42**(12): 821-826.

Javaid, N., F. Haschke, B. Pietschnig, E. Schuster, C. Huemer, A. Shebaz, . . . M. C. Secretin (1991). Interactions between Infections, Malnutrition and Ironnutritional Status in Pakistani Infants - a Longitudinal-Study. *Acta Paediatr Scand* **Suppl374**: 141-150.

JECFA, F. W. (2007). Summary and conclusions of the sixty-eight meeting of Joint FAO/WHO Expert Committee on Food Additives (JECFA). Geneva, Food and Agriculture Organization of the United Nations (FAO) and World Health Organization (WHO).

Kabyemela, E. R., M. Fried, J. D. Kurtis, T. K. Mutabingwa and P. E. Duffy (2008). Decreased susceptibility to Plasmodium falciparum infection in pregnant women with iron deficiency. *J Infect Dis* **198**(2): 163-166.

Kambal, A. E. and E. C. Batesmith (1976). Genetic and Biochemical Study on Pericarp Pigments in a Cross between 2 Cultivars of Grain-Sorghum, Sorghum Bicolor. *Heredity* **37**(Dec): 413-416.

Karamac, M. (2009). Chelation of Cu(II), Zn(II), and Fe(II) by Tannin Constituents of Selected Edible Nuts. *Int J Mol Sci* **10**(12): 5485-5497.

Kartikasari, A. E., N. A. Georgiou, F. L. Visseren, H. van Kats-Renaud, B. S. van Asbeck and J. J. Marx (2006). Endothelial activation and induction of monocyte adhesion by nontransferrin-bound iron present in human sera. *Faseb J* **20**(2): 353-355.

Kawakami, K., S. Aketa, M. Nakanami, S. Iizuka and M. Hirayama (2010). Major Water-Soluble Polyphenols, Proanthocyanidins, in Leaves of Persimmon (Diospyros kaki) and Their alpha-Amylase Inhibitory Activity. *Biosci Biotech Bioch* **74**(7): 1380-1385.

Kayode, A. P., A. R. Linnemann, J. D. Hounhouigan, M. J. Nout and M. A. van Boekel (2006). Genetic and environmental impact on iron, zinc, and phytate in food sorghum grown in Benin. *J Agric Food Chem* **54**(1): 256-262.

Kayode, A. P. P., A. Adegbidi, J. D. Hounhouigan, A. R. Linnemann and M. J. R. Nout (2005). Quality of farmers' varieties of sorghum and derived foods as perceived by consumers in Benin. *Ecol Food Nutr* **44**(4): 271-294.

Kayode, A. P. P., J. D. Hounhouigan and M. J. R. Nout (2007a). Impact of brewing process operations on phytate, phenolic compounds and in vitro solubility of iron and zinc in opaque sorghum beer. *Lwt-Food Sci Technol* **40**(5): 834-841.

Kayode, A. P. P., A. R. Linnemann, M. J. R. Nout and M. A. J. S. Van Boekel (2007b). Impact of sorghum processing on phytate, pnenolic compounds and in vitro solubility of iron and zinc in thick porridges. *J Sci Food Agric* **87**(5): 832-838.

Kennedy, E., V. Mannar and V. Iyengar (2003). Alleviating hidden hunger - Approaches that work. *IAEA Bulletin* **45**(1): 54-60.

Khetarpaul, N. and B. M. Chauhan (1990). Effects of Germination and Pure Culture Fermentation by Yeasts and Lactobacilli on Phytic Acid and Polyphenol Content of Pearl-Millet. *J Food Sci* **55**(4): 1180-&.

Khetarpaul, N. and B. M. Chauhan (1991). Sequential Fermentation of Pearl-Millet by Yeasts and Lactobacilli Effect on the Antinutrients and Invitro Digestibility. *Plant Food Hum Nutr* **41**(4): 321-327.

Khokhar, S. and R. K. O. Apenten (2003). Iron binding characteristics of phenolic compounds: some tentative structure-activity relations. *Food Chem* **81**(1): 133-140.

Khokhar, S., Pushpanjali and G. R. Fenwick (1994). Phytate Content of Indian Foods and Intakes by Vegetarian Indians of Hisar Region, Haryana State. *J Agr Food Chem* **42**(11): 2440-2444.

Kimbi, H. K., D. Nformi and K. J. Ndamukong (2005). Prevalence of asymptomatic malaria among school children in an urban and rural area in the Mount Cameroon region. *Cent Afr J Med* **51**(1-2): 5-10.

Kleih, U., B. S. Ravi, D. B. Rao and B. Yoganand (2007). Industrial Utilization of Sorghum in India. *Journal of SAT agricultural research* **3**(1).

Kodish, S., J. H. Rah, K. Kraemer, S. de Pee and J. Gittelsohn (2011). Understanding low usage of micronutrient powder in the Kakuma Refugee Camp, Kenya: findings from a qualitative study. *Food Nutr Bull* **32**(3): 292-303.

Koreissi, Y., N. Fanou-Fogny, P. J. M. Hulshof and I. D. Brouwer (2012). Fonio (Digitaria exilis) landraces in Mali: Nutrient and phytate content, genetic diversity and effect of processing. *J Food Compos Anal* **in press**.

Kounnavong, S., T. Sunahara, C. G. Mascie-Taylor, M. Hashizume, J. Okumura, K. Moji, . . . T. Yamamoto (2011). Effect of daily versus weekly home fortification with multiple micronutrient powder on haemoglobin concentration of young children in a rural area, Lao People's Democratic Republic: a randomised trial. *Nutr J* **10**: 129.

Krishnan, R., U. Dharmaraj and N. G. Malleshi (2012). Influence of decortication, popping and malting on bioaccessibility of calcium, iron and zinc in finger millet. *Lwt-Food Sci Technol* **48**(2): 169-174.

Krueger, C. G., M. M. Vestling and J. D. Reed (2003). Matrix-assisted laser desorption/ionization time-of-flight mass spectrometry of heteropolyflavan-3-ols and glucosylated heteropolyflavans in sorghum [Sorghum bicolor (L.) Moench]. *J Agr Food Chem* **51**(3): 538-543.

Kruger, J., J. R. N. Taylor and A. Oelofse (2012). Effects of reducing phytate content in sorghum through genetic modification and fermentation on in vitro iron availability in whole grain porridges. *Food Chem* **131**(1): 220-224.

Kudou, S., Y. Fleury, D. Welti, D. Magnolato, T. Uchida, K. Kitamura and K. Okubo (1991). Malonyl Isoflavone Glycosides in Soybean Seeds (Glycine-Max Merrill). *Agr Biol Chem Tokyo* **55**(9): 2227-2233.

Kumar, V., A. K. Sinha, H. P. S. Makkar and K. Becker (2010). Dietary roles of phytate and phytase in human nutrition: A review. *Food Chem* **120**(4): 945-959.

Kurtzhals, J. A., O. Rodrigues, M. Addae, J. O. Commey, F. K. Nkrumah and L. Hviid (1997). Reversible suppression of bone marrow response to erythropoietin in Plasmodium falciparum malaria. *Brit J Haematol* **97**(1): 169-174.

Kuta, D. D., E. Kwon-Ndung, S. Dachi, M. Ukwungwu and E. D. Imolehin (2003). Role of biotechnological interventions in the improvement of "lost crops of Africa": the case of fonio (*Digitaria exilis* and *Digitaria iburua*). *African Journal of Biotechnology* **2**(12): 580-585.

Kuusipalo, H., K. Maleta, A. Briend, M. Manary and P. Ashorn (2006). Growth and change in blood haemoglobin concentration among underweight Malawian infants receiving fortified spreads for 12 weeks: a preliminary trial. *J Pediatr Gastroenterol Nutr* **43**(4): 525-532.

Laishram, D. D., P. L. Sutton, N. Nanda, V. L. Sharma, R. C. Sobti, J. M. Carlton and H. Joshi (2012). The complexities of malaria disease manifestations with a focus on asymptomatic malaria. *Malar J* **11**: 29.

Lamikanra, A. A., D. Brown, A. Potocnik, C. Casals-Pascual, J. Langhorne and D. J. Roberts (2007). Malarial anemia: of mice and men. *Blood* **110**(1): 18-28.

Larson, S. R., J. N. Rutger, K. A. Young and V. Raboy (2000). Isolation and genetic mapping of a non-lethal rice (Oryza sativa L.) low phytic acid 1 mutation. *Crop Sci* **40**(5): 1397-1405.

Larson, S. R., K. A. Young, A. Cook, T. K. Blake and V. Raboy (1998). Linkage mapping of two mutations that reduce phytic acid content of barley grain. *Theor Appl Genet* **97**(1-2): 141-146.

Lartey, A., A. Manu, K. H. Brown, J. M. Peerson and K. G. Dewey (1999). A randomized, community-based trial of the effects of improved, centrally processed complementary foods on growth and micronutrient status of Ghanaian infants from 6 to 12 mo of age. *Am J Clin Nutr* **70**(3): 391-404.

Lawless, J. W., M. C. Latham, L. S. Stephenson, S. N. Kinoti and A. M. Pertet (1994). Iron supplementation improves appetite and growth in anemic Kenyan primary school children. *J Nutr* **124**(5): 645-654.

Layrisse, M., M. N. Garcia-Casal, L. Solano, M. A. Baron, F. Arguello, D. Llovera, . . . E. Tropper (2000). Iron bioavailability in humans from breakfasts enriched with iron bis-glycine chelate, phylates and polyphenols (vol 130, pg 2195, 2000). *J Nutr* **130**(12): 3106-3106.

Lemaire, M., Q. S. Islam, H. Shen, M. A. Khan, M. Parveen, F. Abedin, . . . S. H. Zlotkin (2011). Iron-containing micronutrient powder provided to children with moderate-to-severe malnutrition increases hemoglobin concentrations but not the risk of infectious morbidity: a randomized, double-blind, placebo-controlled, noninferiority safety trial. *Am J Clin Nutr* **94**(2): 585-593.

Lempereur, I., X. Rouau and J. Abecassis (1997). Genetic and agronomic variation in arabinoxylan and ferulic acid contents of durum wheat (Triticum durum L.) grain and its milling fractions. *J Cereal Sci* **25**(2): 103-110.

Lestienne, I., M. Buisson, V. Lullien-Pellerin, C. Picq and S. Treche (2007). Losses of nutrients and anti-nutritional factors during abrasive decortication of two pearl millet cultivars (Pennisetum glaucum). *Food Chem* **100**(4): 1316-1323.

Lestienne, I., B. Caporiccio, P. Bresancon, I. Rochette and S. Trechet (2005). Relative contribution of phytates, fibers, and tannins to low iron and zinc in vitro solubility in pearl millet (Pennisetum glaucum) flour and grain fractions. *J Agr Food Chem* **53**(21): 8342-8348.

Lin, C. A., M. J. Manary, K. Maleta, A. Briend and P. Ashorn (2008). An energy-dense complementary food is associated with a modest increase in weight gain when compared with a fortified porridge in Malawian children aged 6-18 months. *J Nutr* **138**(3): 593-598.

Liyanage, C. and S. Zlotkin (2002). Bioavailability of iron from micro-encapsulated iron sprinkle supplement. *Food Nutr Bull* **23**(3 Suppl): 133-137.

Longfils, P., D. Monchy, H. Weinheimer, V. Chavasit, Y. Nakanishi and K. Schumann (2008). A comparative intervention trial on fish sauce fortified with NaFe-EDTA and FeSO4+citrate in iron deficiency anemic school children in Kampot, Cambodia. *Asia Pac J Clin Nutr* **17**(2): 250-257.

Lonnerdal, B. (2010). Alternative pathways for absorption of iron from foods *Pure Appl Chem* **82**(2): 429-436.

Lopez de Romana, D., F. Pizarro, D. Diazgranados, A. Barba, M. Olivares and O. Brunser (2011). Effect of Helicobacter pylori infection on iron absorption in asymptomatic adults consuming wheat flour fortified with iron and zinc. *Biol Trace Elem Res* **144**(1-3): 1318-1326.

Loyevsky, M., C. John, B. Dickens, V. Hu, J. H. Miller and V. R. Gordeuk (1999). Chelation of iron within the erythrocytic Plasmodium falciparum parasite by iron chelators. *Mol Biochem Parasitol* **101**(1-2): 43-59.

Lozoff, B. (2007). Iron deficiency and child development. *Food Nutr Bull* **28**(4 Suppl): S560-571.

Lucca, P., R. Hurrell and I. Potrykus (2001). Genetic engineering approaches to improve the bioavailability and the level of iron in rice grains. *Theor Appl Genet* **102**(2-3): 392-397.

Lukac, R. J., M. R. Aluru and M. B. Reddy (2009). Quantification of Ferritin from Staple Food Crops. *J Agr Food Chem* **57**(6): 2155-2161.

Lundeen, E., T. Schueth, N. Toktobaev, S. Zlotkin, S. M. Hyder and R. Houser (2010). Daily use of Sprinkles micronutrient powder for 2 months reduces anemia among children 6 to 36 months of age in the Kyrgyz Republic: a cluster-randomized trial. *Food Nutr Bull* **31**(3): 446-460.

Luten, J., H. Crews, A. Flynn, P. VanDael, P. Kastenmayer, R. Hurrell, . . . W. Frohlich (1996). Interlaboratory trial on the determination of the in vitro iron dialysability from food. *J Sci Food Agric* **72**(4): 415-424.

Lutter, C. K., A. Rodriguez, G. Fuenmayor, L. Avila, F. Sempertegui and J. Escobar (2008). Growth and micronutrient status in children receiving a fortified complementary food. *J Nutr* **138**(2): 379-388.

Lyke, K. E., R. Burges, Y. Cissoko, L. Sangare, M. Dao, I. Diarra, . . . M. B. Sztein (2004). Serum Levels of the Proinflammatory Cytokines Interleukin-1 Beta (IL-1{beta}), IL-6, IL-8, IL-10, Tumor Necrosis Factor Alpha, and IL-12(p70) in Malian Children with Severe Plasmodium falciparum Malaria and Matched Uncomplicated Malaria or Healthy Controls. *Infect Immun* **72**(10): 5630-5637.

Lykke, A. M., O. Mertz and S. Ganaba (2002). Food consumption in rural Burkina Faso. *Ecol Food Nutr* **41**(2): 119-153.

Lynch, S. R., T. Bothwel and S. T. F. I. Powders (2007). A comparison of physical properties, screening procedures and a human efficacy trial for predicting the bioavailability of commercial elemental iron powders used for food fortification. *Int J Vitam Nutr Res* **77**(2): 107-124.

Lynch, S. R., S. A. Dassenko, J. D. Cook, M. A. Juillerat and R. F. Hurrell (1994). Inhibitory effect of a soybean-protein--related moiety on iron absorption in humans. *Am J Clin Nutr* **60**(4): 567-572.

Macharia-Mutie, C. W., D. Moretti, N. Van den Briel, A. M. Omusundi, A. M. Mwangi, F. J. Kok, . . . I. D. Brouwer (2012). Maize porridge enriched with a micronutrient powder containing low-dose iron as NaFeEDTA but not amaranth grain flour reduces anemia and iron deficiency in Kenyan preschool children. *J Nutr* **142**(9): 1756-1763.

Macphail, A. P., T. H. Bothwell, J. D. Torrance, D. P. Derman, W. R. Bezwoda, R. W. Charlton and F. Mayet (1981). Factors Affecting the Absorption of Iron from Fe(III)EDTA. *Brit J Nutr* **45**(2): 215-227.

Macphail, A. P., R. C. Patel, T. H. Bothwell and R. D. Lamparelli (1994). EDTA and the Absorption of Iron from Food. *Am J Clin Nutr* **59**(3): 644-648.

Mahajan, S. and B. M. Chauhan (1987). Phytic Acid and Extractable Phosphorus of Pearl-Millet Flour as Affected by Natural Lactic-Acid Fermentation. *J Sci Food Agric* **41**(4): 381-386.

Mahgoub, S. E. O. and S. A. Elhag (1998). Effect of milling, soaking, malting, heat treatment and fermentation on phytate level of four Sudanese sorghum cultivars. *Food Chem* **61**(1-2): 77-80.

Makokha, A. O., R. K. Oniang'o, S. M. Njoroge and O. K. Kamar (2002). Effect of traditional fermentation and malting on phytic acid and mineral availability from sorghum (Sorghum bicolor) and finger millet (Eleusine coracana) grain varieties grown in Kenya. *Food Nutr Bull* **23**(3 Suppl): 241-245.

Malone, H. E., J. P. Kevany, J. M. Scott, S. D. O'Broin and G. O'Connor (1986). Ascorbic acid supplementation: its effects on body iron stores and white blood cells. *Ir J Med Sci* **155**(3): 74-79.

Manach, C., A. Scalbert, C. Morand, C. Remesy and L. Jimenez (2004). Polyphenols: food sources and bioavailability. *Am J Clin Nutr* **79**(5): 727-747.

Mao, X. and G. Yao (1992). Effect of vitamin C supplementations on iron deficiency anemia in Chinese children. *Biomed Environ Sci* **5**(2): 125-129.

Matuschek, E., E. Towo and U. Svanberg (2001). Oxidation of polyphenols in phytate-reduced high-tannin cereals: Effect on different phenolic groups and on in vitro accessible iron. *J Agr Food Chem* **49**(11): 5630-5638.

Maxson, E. D., L. W. Rooney, J. W. Johnson and L. E. Clark (1972). Factors Affecting Tannin Content of Sorghum Grain as Determined by 2 Methods of Tannin Analysis. *Crop Science* **12**(2): 233-&.

Mazza, G. (2007). Bioactivity, absorption and metabolism of Anthocyanins. *Acta Hortic*(744): 117-125.
Mbithi-Mwikya, S., J. Van Camp, Y. Yiru and A. Huyghebaert (2000). Nutrient and antinutrient changes in finger millet (Eleusine coracan) during sprouting. *Lwt-Food Sci Technol* **33**(1): 9-14.
McDonough, C. M. and L. W. Rooney (2000) The millets. In *Handbook of cereal science and technology*, pp. 177-201 [K. Kulp and J. G. Ponte jr., editors]. New York: Marcel Dekker Inc.
Mebrahtu, T., R. J. Stoltzfus, H. M. Chwaya, J. K. Jape, L. Savioli, A. Montresor, . . . J. M. Tielsch (2004). Low-dose daily iron supplementation for 12 months does not increase the prevalence of malarial infection or density of parasites in young Zanzibari children. *J Nutr* **134**(11): 3037-3041.
Mendoza, C., F. E. Viteri, B. Lonnerdal, V. Raboy, K. A. Young and K. H. Brown (2001). Absorption of iron from unmodified maize and genetically altered, low-phytate maize fortified with ferrous sulfate or sodium iron EDTA. *Am J Clin Nutr* **73**(1): 80-85.
Mendoza, C., F. E. Viteri, B. Lonnerdal, K. A. Young, V. Raboy and K. H. Brown (1998). Effect of genetically modified, low-phytic acid maize on absorption of iron from tortillas. *Am J Clin Nutr* **68**(5): 1123-1127.
Menendez, C., E. Kahigwa, R. Hirt, P. Vounatsou, J. J. Aponte, F. Font, . . . P. L. Alonso (1997). Randomised placebo-controlled trial of iron supplementation and malaria chemoprophylaxis for prevention of severe anaemia and malaria in Tanzanian infants. *Lancet* **350**(9081): 844-850.
Menendez, C., J. Todd, P. L. Alonso, N. Francis, S. Lulat, S. Ceesay, . . . B. M. Greenwood (1994). The effects of iron supplementation during pregnancy, given by traditional birth attendants, on the prevalence of anaemia and malaria. *Trans R Soc Trop Med Hyg* **88**(5): 590-593.
Menon, P., M. T. Ruel, C. U. Loechl, M. Arimond, J. P. Habicht, G. Pelto and L. Michaud (2007). Micronutrient Sprinkles reduce anemia among 9- to 24-mo-old children when delivered through an integrated health and nutrition program in rural Haiti. *J Nutr* **137**(4): 1023-1030.
Merhav, H., Y. Amitai, H. Palti and S. Godfrey (1985). Tea Drinking and Microcytic Anemia in Infants. *Am J Clin Nutr* **41**(6): 1210-1213.
Michaelsen, K. F., C. Hoppe, N. Roos, P. Kaestel, M. Stougaard, L. Lauritzen, H. Friis (2009). Choice of foods and ingredients for moderately malnourished children 6 months to 5 years of age. *Food Nutr Bull* **30**(3): S343-S404.
Michalakis, Y. and F. Renaud (2009). Malaria: Evolution in vector control. *Nature* **462**(7271): 298-300.
Miglioranza, L. H. S., T. Matsuo, G. M. Caballero-Cordoba, J. B. Dichi, E. S. Cyrino, I. B. N. Oliveira, . . . I. Dichi (2003). Effect of long-term fortification of whey drink with ferrous bisglycinate on anemia prevalence in children and adolescents from deprived areas in Londrina, Parana, Brazil. *Nutrition* **19**(5): 419-421.
Miller, D. D., B. R. Schricker, R. R. Rasmussen and D. Vancampen (1981). An Invitro Method for Estimation of Iron Availability from Meals. *Am J Clin Nutr* **34**(10): 2248-2256.
Miller, L. H., D. I. Baruch, K. Marsh and O. K. Doumbo (2002). The pathogenic basis of malaria. *Nature* **415**(6872): 673-679.
Miller, L. H., M. F. Good and G. Milon (1994). Malaria pathogenesis. *Science* **264**(5167): 1878-1883.
Mira, L., M. T. Fernandez, M. Santos, R. Rocha, M. H. Florencio and K. R. Jennings (2002). Interactions of flavonoids with iron and copper ions: A mechanism for their antioxidant activity. *Free Radical Res* **36**(11): 1199-1208.
Moffatt, M. E., S. Longstaffe, J. Besant and C. Dureski (1994). Prevention of iron deficiency and psychomotor decline in high-risk infants through use of iron-fortified infant formula: a randomized clinical trial. *J Pediatr* **125**(4): 527-534.

Monsen, E. R. and J. D. Cook (1976). Food iron absorption in human subjects. IV. The effects of calcium and phosphate salts on the absorption of nonheme iron. *Am J Clin Nutr* **29**(10): 1142-1148.

Monsen, E. R., L. Hallberg, M. Layrisse, D. M. Hegsted, J. D. Cook, W. Mertz and C. A. Finch (1978). Estimation of available dietary iron. *Am J Clin Nutr* **31**(1): 134-141.

Moretti, D., M. B. Zimmermann, S. Muthayya, P. Thankachan, T. C. Lee, A. V. Kurpad and R. F. Hurrell (2006a). Extruded rice fortified with micronized ground ferric pyrophosphate reduces iron deficiency in Indian schoolchildren: a double-blind randomized controlled trial. *Am J Clin Nutr* **84**(4): 822-829.

Moretti, D., M. B. Zimmermann, R. Wegmuller, T. Walczyk, C. Zeder and R. F. Hurrell (2006b). Iron status and food matrix strongly affect the relative bioavailability of ferric pyrophosphate in humans. *Am J Clin Nutr* **83**(3): 632-638.

Mouquet-Rivier, C., C. Icard-Verniere, J. P. Guyot, E. Hassane Tou, I. Rochette and S. Treche (2008). Consumption pattern, biochemical composition and nutritional value of fermented pearl millet gruels in Burkina Faso. *Int J Food Sci Nutr* **59**(7-8): 716-729.

Murphy, S. P., D. H. Calloway and G. H. Beaton (1995). Schoolchildren Have Similar Predicted Prevalences of Inadequate Intakes as Toddlers in Village Populations in Egypt, Kenya, and Mexico. *Eur J Clin Nutr* **49**(9): 647-657.

Murray, C. J., L. C. Rosenfeld, S. S. Lim, K. G. Andrews, K. J. Foreman, D. Haring, . . . A. D. Lopez (2012). Global malaria mortality between 1980 and 2010: a systematic analysis. *Lancet* **379**(9814): 413-431.

Murray, M. J., A. B. Murray, M. B. Murray and C. J. Murray (1978). Adverse Effect of Iron Repletion on Course of Certain Infections. *Brit Med J* **2**(6145): 1113-1115.

Murty, D. S. and V. Subramanian (2009) Sorghum Roti: I. traditional methods of consumption and standard procedures for evaluation. http://oar.icrisat.org/4091/1/CP_90.pdf (accessed November 2012).

Muthayya, S., P. Thankachan, S. Hirve, V. Amalrajan, T. Thomas, H. Lubree, . . . A. V. Kurpad (2012). Iron fortification of whole wheat flour reduces iron deficiency and iron deficiency anemia and increases body iron stores in Indian school-aged children. *J Nutr* **142**(11): 1997-2003.

Mwesigye, P. K. and T. O. Okurut (1995). A Survey of the Production and Consumption of Traditional Alcoholic Beverages in Uganda. *Process Biochem* **30**(6): 497-501.

Naczk, M., R. Amarowicz, R. Zadernowski and F. Shahidi (2001). Protein-precipitating capacity of crude condensed tannins of canola and rapeseed hulls. *J Am Oil Chem Soc* **78**(12): 1173-1178.

Nambiar, V. S., J. J. Dhaduk, N. Sareen, T. Shahu and D. R. (2011). Potential functional implications of pearl millet (*Pennisetum glaucum*) in health and disease. *Journal of Applied Pharmaceutical Science* **1**(10): 62-67.

Navert, B., B. Sandstrom and A. Cederblad (1985). Reduction of the Phytate Content of Bran by Leavening in Bread and Its Effect on Zinc-Absorption in Man. *Brit J Nutr* **53**(1): 47-53.

Ndemwa, P., C. L. Klotz, D. Mwaniki, K. Sun, E. Muniu, P. Andango, . . . R. D. Semba (2011). Relationship of the availability of micronutrient powder with iron status and hemoglobin among women and children in the Kakuma Refugee Camp, Kenya. *Food Nutr Bull* **32**(3): 286-291.

Ndjeunga, J. and C. H. Nelson (1999) Prospects for a Pearl Millet and Sorghum Food Processing Industry in West Africa Semi-Arid Tropics. In: Towards Sustainable Sorghum Production, Utilization, and Commercialization in West and Central Africa Proceedings of a Technical Workshop of the West and Central Africa Sorghum Research Network. http://oar.icrisat.org/4891/ (accessed January 2013).

Nemeth, E. and T. Ganz (2009). The role of hepcidin in iron metabolism. *Acta Haematol* **122**(2-3): 78-86.

Nemeth, E., M. S. Tuttle, J. Powelson, M. B. Vaughn, A. Donovan, D. M. Ward, . . . J. Kaplan (2004). Hepcidin regulates cellular iron efflux by binding to ferroportin and inducing its internalization. *Science* **306**(5704): 2090-2093.

Nestel, P. and D. Alnwick (1996) Iron/multi-micronutrient supplements for young children: summary and conclusions of a consultation held at UNICEF. http://pdf.usaid.gov/pdf_docs/PNACM315.pdf (accessed November 2012).

Nestel, P., H. E. Bouis, J. V. Meenakshi and W. Pfeiffer (2006). Biofortification of staple food crops. *J Nutr* **136**(4): 1064-1067.

Nestel, P., R. Nalubola, R. Sivakaneshan, A. R. Wickramasinghe, S. Atukorala, T. Wickramanayake and F. F. T. Team (2004). The use of iron-fortified wheat flour to reduce anemia among the estate population in Sri Lanka. *Int J Vitam Nutr Res* **74**(1): 35-51.

Netherwood, T., S. M. Martin-Orue, A. G. O'Donnell, S. Gockling, J. Graham, J. C. Mathers and H. J. Gilbert (2004). Assessing the survival of transgenic plant DNA in the human gastrointestinal tract. *Nat Biotechnol* **22**(2): 204-209.

Newton, C. R., P. A. Warn, P. A. Winstanley, N. Peshu, R. W. Snow, G. Pasvol and K. Marsh (1997). Severe anaemia in children living in a malaria endemic area of Kenya. *Trop Med Int Health* **2**(2): 165-178.

Nip, W. K. and E. E. Burns (1969). Pigment Characterization in Grain Sorghum .I. Red Varieties. *Cereal Chem* **46**(5): 490-&.

Nout, M. J. R. (2009). Rich nutrition from the poorest - Cereal fermentations in Africa and Asia. *Food Microbiol* **26**(7): 685-692.

Nwagha, U. I., V. O. Ugwu, T. U. Nwagha and B. U. Anyaehie (2009). Asymptomatic Plasmodium parasitaemia in pregnant Nigerian women: almost a decade after Roll Back Malaria. *Trans R Soc Trop Med Hyg* **103**(1): 16-20.

Nwanyanwu, O. C., C. Ziba, P. N. Kazembe, G. Gamadzi, J. Gandwe and S. C. Redd (1996). The effect of oral iron therapy during treatment for Plasmodium falciparum malaria with sulphadoxine-pyrimethamine on Malawian children under 5 years of age. *Ann Trop Med Parasitol* **90**(6): 589-595.

Nyakeriga, A. M., M. Troye-Blomberg, J. R. Dorfman, N. D. Alexander, R. Back, M. Kortok, . . . T. N. Williams (2004). Iron deficiency and malaria among children living on the coast of Kenya. *J Infect Dis* **190**(3): 439-447.

O'Donnell, A., F. J. I. Fowkes, S. J. Allen, H. Imrie, M. P. Alpers, D. I. Weatherall and K. P. Day (2009). The acute phase response in children with mild and severe malaria in Papua New Guinea. *Trans R Soc Trop Med Hyg* **103**(7): 679-686.

Obilana, A. B. (2003) Overview: importance of millets in Africa. Afripro, Workshop on the proteins of sorghum and millets: Enhancing nutritional and functional properties for Africa. Pretoria, South Africa, 2-4 April 2003. http://www.afripro.org.uk/papers/paper02Obilana.pdf (accessed January 2013).

Odell, B. L., A. R. Deboland and Koirtyoh.Sr (1972). Distribution of Phytate and Nutritionally Important Elements among Morphological Components of Cereal Grains. *J Agr Food Chem* **20**(3): 718-&.

Oghbaei, M. and J. Prakash (2012). Bioaccessible nutrients and bioactive components from fortified products prepared using finger millet (Eleusine coracana). *J Sci Food Agric* **92**(11): 2281-2290.

Ojukwu, J. U., J. U. Okebe, D. Yahav and M. Paul (2009). Oral iron supplementation for preventing or treating anaemia among children in malaria-endemic areas. *Cochrane Database Syst Rev*(3): CD006589.

Okebe, J. U., D. Yahav, R. Shbita and M. Paul (2011). Oral iron supplements for children in malaria-endemic areas. *Cochrane Database Syst Rev*(10): CD006589.

Olewnik, M. C., R. C. Hoseney and E. Varrianomarston (1984). A Procedure to Produce Pearl-Millet Rotis. *Cereal Chem* **61**(1): 28-33.

Onofiok, N. O. and D. O. Nnanyelugo (1998). Weaning foods in West Africa: Nutritional problems and possible solutions. *Food Nutr Bull* **19**.

Onyenekwe, C. C., S. C. Meludu, C. E. Dioka and L. S. Salimonu (2002). Prevalence of asymptomatic malaria parasitaemia amongst pregnant women. *Indian J Malariol* **39**(3-4): 60-65.

Oppenheimer, S. J., F. D. Gibson, S. B. Macfarlane, J. B. Moody, C. Harrison, A. Spencer and O. Bunari (1986). Iron supplementation increases prevalence and effects of malaria: report on clinical studies in Papua New Guinea. *Trans R Soc Trop Med Hyg* **80**(4): 603-612.

Ortiz-Monasterio, J. I., N. Palacios-Rojas, E. Meng, K. Pixley, R. Trethowan and R. J. Pena (2007). Enhancing the mineral and vitamin content of wheat and maize through plant breeding. *J Cereal Sci* **46**(3): 293-307.

Osman, A. K. and A. Al-Othaimeen (2002). Experience with ferrous bis-glycine chelate as an iron fortificant in milk. *Int J Vitam Nutr Res* **72**(4): 257-263.

Osman, M. A. (2004). Changes in sorghum enzyme inhibitors, phytic acid, tannins and in vitro protein digestibility occurring during Khamir (local bread) fermentation. *Food Chem* **88**(1): 129-134.

Oury, F. X., F. Leenhardt, C. Remesy, E. Chanliaud, B. Duperrier, F. Balfourier and G. Charmet (2006). Genetic variability and stability of grain magnesium, zinc and iron concentrations in bread wheat. *Eur J Agron* **25**(2): 177-185.

Pale, S., S. J. B. Taonda, B. Bougouma and S. C. Mason (2010). Sorghum Malt and Traditional Beer (Dolo) Quality Assessment in Burkina Faso. *Ecol Food Nutr* **49**(2): 129-141.

Pardey, P. G., B. D. Wright and C. Nottenburg (2001) Are intellectual property rights stifling agricultural biotechnology in developing countries http://www.ifpri.org/publication/are-intellectual-property-rights-stifling-agricultural-biotechnology-developing-countrie (accessed November 2012).

Parthasarathy Rao, P., P. S. Birthal, B. V. S. Reddy, K. N. Rai and S. Ramesh (2006). Diagnostics of Sorghum and Pearl Millet Grains-based Nutrition in India. *Journal of SAT agricultural research* **2**(1).

Pawar, V. D. and G. S. Parlikar (1990). Reducing the Polyphenols and Phytate and Improving the Protein-Quality of Pearl-Millet by Dehulling and Soaking. *J Food Sci Tech Mys* **27**(3): 140-143.

Penalvo, J. L., K. M. Haajanen, N. Botting and H. Adlercreutz (2005). Quantification of lignans in food using isotope dilution gas chromatography/mass spectrometry. *J Agr Food Chem* **53**(24): 9342-9347.

Perez-Exposito, A. B., S. Villalpando, J. A. Rivera, I. J. Griffin and S. A. Abrams (2005). Ferrous sulfate is more bioavailable among preschoolers than other forms of iron in a milk-based weaning food distributed by PROGRESA, a national program in Mexico. *J Nutr* **135**(1): 64-69.

Perron, N. R. and J. L. Brumaghim (2009). A Review of the Antioxidant Mechanisms of Polyphenol Compounds Related to Iron Binding. *Cell Biochem Biophys* **53**(2): 75-100.

Peto, T. E. and J. L. Thompson (1986). A reappraisal of the effects of iron and desferrioxamine on the growth of Plasmodium falciparum 'in vitro': the unimportance of serum iron. *Br. J. Haematol.* **63**(2): 273-280.

Petry, N., I. Egli, J. B. Gahutu, P. L. Tugirimana, E. Boy and R. Hurrell (2012). Stable Iron Isotope Studies in Rwandese Women Indicate That the Common Bean Has Limited Potential as a Vehicle for Iron Biofortification. *J Nutr* **142**(3): 492-497.

Petry, N., I. Egli, C. Zeder, T. Walczyk and R. Hurrell (2010). Polyphenols and phytic acid contribute to the low iron bioavailability from common beans in young women. *J Nutr* **140**(11): 1977-1982.

Pfeiffer, W. and B. McClafferty (2007) Chapter 3. Biofortification: breeding micronutrient-dense crops. In *Breeding major food staples*, 4th ed., [M. S. Kang and P. M. Priyadarshan, editors]. Oxford, UK: Wiley-Blackwell Publishing.

Phuka, J., U. Ashorn, P. Ashorn, M. Zeilani, Y. B. Cheung, K. G. Dewey, . . . K. Maleta (2011). Acceptability of three novel lipid-based nutrient supplements among Malawian infants and their caregivers. *Matern Child Nutr* **7**(4): 368-377.

Phuka, J. C., K. Maleta, C. Thakwalakwa, Y. B. Cheung, A. Briend, M. J. Manary and P. Ashorn (2009). Postintervention growth of Malawian children who received 12-mo dietary complementation with a lipid-based nutrient supplement or maize-soy flour. *Am J Clin Nutr* **89**(1): 382-390.

Piccinin, D. (2002) More about Ethiopian Food: tef. http://bsesrv214.bse.vt.edu/Grisso/Ethiopia/Books_Resources/Threshing/teff.pdf

Prentice, A. M. (2008). Iron Metabolism, Malaria, and Other Infections: What Is All the Fuss About? *J Nutr* **138**(12): 2537-2541.

Prentice, A. M. and S. E. Cox (2012). Iron and malaria interactions: research needs from basic science to global policy. *Adv Nutr* **3**(4): 583-591.

Prentice, A. M., H. Ghattas, C. Doherty and S. E. Cox (2007). Iron metabolism and malaria. *Food Nutr Bull* **28**(4 Suppl): S524-539.

Price, M. L. and L. G. Butler (1977). Rapid Visual Estimation and Spectrophotometric Determination of Tannin Content of Sorghum Grain. *J Agr Food Chem* **25**(6): 1268-1273.

Purawatt, S., A. Siripinyanond and J. Shiowatana (2007). Flow field-flow fractionation-inductively coupled optical emission spectrometric investigation of the size-based distribution of iron complexed to phytic and tannic acids in a food suspension: implications for iron availability. *Anal Bioanal Chem* **389**(3): 733-742.

Qaim, M., A. J. Stein and J. V. Meenakshi (2007). Economics of biofortification. *Agr Econ-Blackwell* **37**: 119-133.

Raboy, V. (1997) *Accumulation and storage of phosphate and minerals*. Dordrecht: Kluwer Academic Publishers.

Raboy, V. (2001). Seeds for a better future: 'low phytate', grains help to overcome malnutrition and reduce pollution. *Trends Plant Sci* **6**(10): 458-462.

Raboy, V. (2003). myo-Inositol-1,2,3,4,5,6-hexakisphosphate. *Phytochemistry* **64**(6): 1033-1043.

Raboy, V., P. F. Gerbasi, K. A. Young, S. D. Stoneberg, S. G. Pickett, A. T. Bauman, . . . D. S. Ertl (2000). Origin and seed phenotype of maize low phytic acid 1-1 and low phytic acid 2-1. *Plant Physiol.* **124**(1): 355-368.

Radhakrishnan, M. R. and J. Sivaprasad (1980). Tannin Content of Sorghum Varieties and Their Role in Iron Bioavailability. *J Agr Food Chem* **28**(1): 55-57.

Ragaee, S., E. M. Abdel-Aal and M. Noaman (2006). Antioxidant activity and nutrient composition of selected cereals for food use. *Food Chem* **98**(1): 32-38.

Rahman, I. E. A. and M. A. W. Osman (2011). Effect of sorghum type (Sorghum bicolor) and traditional fermentation on tannins and phytic acid contents and trypsin inhibitor activity. *J Food Agric Environ* **9**(3-4): 163-166.

Rai, K. N., M. Govindaraj and A. S. Rao (2012). Genetic enhancement of grain iron and zinc content in pearl millet. *Qual Assur Saf Crop* **4**(3): 119-125.

Ramachandra, G., T. K. Virupaksha and M. Shadaksharaswamy (1977). Relationship between tannin levels and in vitro protein digestibility in finger millet (Eleusine coracana Gaertn.). *J Agric Food Chem* **25**(5): 1101-1104.

Ranum, P. and A. Wesley (2008) *Cereal fortification handbook*. Ottawa: Micronutrient Initiative.

Rao, M. V. S. S. T. S. and G. Muralikrishna (2002). Evaluation of the antioxidant properties of free and bound phenolic acids from native and malted finger millet (ragi, Eleusine coracana Indaf-15). *J Agr Food Chem* **50**(4): 889-892.

Rao, N. B. S. (1985) Salt. In *Iron fortification of foods*, pp. 39-53 [F. M. Clydesdale and K. L. Wiemer, editors]. Orlando: Academic Press.

Rao, P. U. and Y. G. Deosthale (1988). Invitro Availability of Iron and Zinc in White and Colored Ragi (Eleusine-Coracana) - Role of Tannin and Phytate. *Plant Food Hum Nutr* **38**(1): 35-41.

Ravindran, V., G. Ravindran and S. Sivalogan (1994). Total and Phytate Phosphorus Contents of Various Foods and Feedstuffs of Plant-Origin. *Food Chem* **50**(2): 133-136.

Reddy, B. V. S., S. Ramesh and T. Longvah (2005). Prospects of breeding for micronutrient and ß-carotene-dense sorghums. *International Sorghum and Millet Newsletter* **47**: 93-96.

Reddy, M. B. and J. D. Cook (1991). Assessment of dietary determinants of nonheme-iron absorption in humans and rats. *Am J Clin Nutr* **54**(4): 723-728.

Reddy, N. R. (2002). In *Food Phytates*, [N. R. Reddy and S. K. Sathe, editors]. Boca Raton, London, New York, Washington DC: CRC Press.

Reichert, R. D. (1979). Ph-Sensitive Pigments in Pearl-Millet. *Cereal Chem* **56**(4): 291-294.

Richard, S. A., N. Zavaleta, L. E. Caulfield, R. E. Black, R. S. Witzig and A. H. Shankar (2006). Zinc and iron supplementation and malaria, diarrhea, and respiratory infections in children in the Peruvian Amazon. *Am J Trop Med Hyg* **75**(1): 126-132.

Rim, H., S. Kim, B. Sim, H. Gang, H. Kim, Y. Kim, . . . M. Yang (2008). Effect of iron fortification of nursery complementary food on iron status of infants in the DPRKorea. *Asia Pac J Clin Nutr* **17**(2): 264-269.

Rivera, J. A., T. Shamah, S. Villalpando and E. Monterrubio (2010). Effectiveness of a large-scale iron-fortified milk distribution program on anemia and iron deficiency in low-income young children in Mexico. *Am J Clin Nutr* **91**(2): 431-439.

Roe, M. A., R. Collings, J. Hoogewerff and S. J. Fairweather-Tait (2009). Relative bioavailability of micronized, dispersible ferric pyrophosphate added to an apple juice drink. *Eur J Nutr* **48**(2): 115-119.

Rohner, F., F. O. Ernst, M. Arnold, M. Hibe, R. Biebinger, F. Ehrensperger, . . . M. B. Zimmermann (2007). Synthesis, characterization, and bioavailability in rats of ferric phosphate nanoparticles. *J Nutr* **137**(3): 614-619.

Rohner, F., M. B. Zimmermann, R. J. Amon, P. Vounatsou, A. B. Tschannen, E. K. N'Goran, . . . R. F. Hurrell (2010). In a Randomized Controlled Trial of Iron Fortification, Anthelmintic Treatment, and Intermittent Preventive Treatment of Malaria for Anemia Control in Ivorian Children, only Anthelmintic Treatment Shows Modest Benefit. *J Nutr* **140**(3): 635-641.

Rooney, L. W. (1978). Sorghum and Pearl Millet Lipids. *Cereal Chem* **55**(5): 584-590.

Rooney, L. W. (2000) Genetics and cytogenetics. In *Sorghum: origin, history, technology and production*, 1st ed., pp. 261-307 [C. W. Smith and R. A. Frederiksen, editors]. New York: John Wiley and Sons.

Rooney, L. W., A. W. Kirleis and D. S. Murty (1986). Traditional foods from sorghum: their production, evaluation and nutritional value. *Adv Cereal Sci Technol* **8**: 317-353.

Rossander-Hulthén, L. and L. Hallberg (1996) Dietary factors influencing iron absorption-an overview. In *Iron nutrition in health and disease*, [L. Hallberg, editor]. London: John Libbey & Company.

Roughead, Z. K., C. A. Zito and J. R. Hunt (2005). Inhibitory effects of dietary calcium on the initial uptake and subsequent retention of heme and nonheme iron in humans: comparisons using an intestinal lavage method. *Am J Clin Nutr* **82**(3): 589-597.

Ruel, M. T. and C. E. Levin (2001) Food-based approaches. In *Nutritional anemias*. Boca Raton (FL): CRC Press.

Samadpour, K., K. Z. Long, R. Hayatbakhsh and G. C. Marks (2011). Randomised comparison of the effects of Sprinkles and Foodlets with the currently recommended supplement (Drops) on micronutrient status and growth in Iranian children. *Eur J Clin Nutr* **65**(12): 1287-1294.

Samman, S., B. Sandstrom, M. B. Toft, K. Bukhave, M. Jensen, S. S. Sorensen and M. Hansen (2001). Green tea or rosemary extract added to foods reduces nonheme-iron absorption. *Am J Clin Nutr* **73**(3): 607-612.

Sanchez-Lopez, R. and K. Haldar (1992). A transferrin-independent iron uptake activity in Plasmodium falciparum-infected and uninfected erythrocytes. *Mol Biochem Parasitol* **55**(1-2): 9-20.

Sandberg, A. S. (2005). Methods and options for in vitro dialyzability; Benefits and limitations. *Int J Vitam Nutr Res* **75**(6): 395-404.

Sandberg, A. S. (2010). The Use of Caco-2 Cells to Estimate Fe Absorption in Humans - a Critical Appraisal. *Int J Vitam Nutr Res* **80**(4-5): 307-313.

Sandberg, A. S., L. R. Hulthen and M. Turk (1996). Dietary Aspergillus niger phytase increases iron absorption in humans. *J Nutr* **126**(2): 476-480.

Sanders, J. H. and B. Ouendeba (2012). Intensive Production of Millet and Sorghum for Evolving Markets in the Sahel. Lincoln, NE, University of Nebraska.

Sarker, S. A., L. Davidsson, H. Mahmud, T. Walczyk, R. F. Hurrell, N. Gyr and G. J. Fuchs (2004). Helicobacter pylori infection, iron absorption, and gastric acid secretion in Bangladeshi children. *Am J Clin Nutr* **80**(1): 149-153.

Sartelet, H., S. Serghat, A. Lobstein, Y. Ingenbleek, R. Anton, E. Petitfrere, . . . B. Haye (1996). Flavonoids extracted from Fonio millet (Digitaria exilis) reveal potent antithyroid properties. *Nutrition* **12**(2): 100-106.

Sazawal, S., R. E. Black, M. Ramsan, H. M. Chwaya, R. J. Stoltzfus, A. Dutta, . . . F. M. Kabole (2006). Effects of routine prophylactic supplementation with iron and folic acid on admission to hospital and mortality in preschool children in a high malaria transmission setting: community-based, randomised, placebo-controlled trial. *Lancet* **367**(9505): 133-143.

Schauer, C. and S. Zlotkin (2003). Home fortification with micronutrient sprinkles - A new approach for the prevention and treatment of nutritional anemias. *Paediatr Child Health* **8**(2): 87-90.

Schlemmer, U., W. Frolich, R. M. Prieto and F. Grases (2009). Phytate in foods and significance for humans: food sources, intake, processing, bioavailability, protective role and analysis. *Mol Nutr Food Res* **53 Suppl 2**: S330-375.

Schonfeldt, H. C. and N. Gibson Hall (2012). Dietary protein quality and malnutrition in Africa. *Br J Nutr* **108 Suppl 2**: S69-76.

Schumann, K., N. W. Solomons, M. E. Romero-Abal, M. Orozco, G. Weiss and J. Marx (2012). Oral administration of ferrous sulfate, but not of iron polymaltose or sodium iron ethylenediaminetetraacetic acid (NaFeEDTA), results in a substantial increase of non-transferrin-bound iron in healthy iron-adequate men. *Food Nutr Bull* **33**(2): 128-136.

Seal, A., E. Kafwembe, I. A. R. Kassim, M. Hong, A. Wesley, J. Wood, . . . T. van den Briel (2008). Maize meal fortification is associated with improved vitamin A and iron status in adolescents and reduced childhood anaemia in a food aid-dependent refugee population. *Public Health Nutr* **11**(7): 720-728.

Senga, E. L., G. Harper, G. Koshy, P. N. Kazembe and B. J. Brabin (2011). Reduced risk for placental malaria in iron deficient women. *Malar J* **10**: 47.

Seralini, G. E., D. Cellier and J. S. de Vendomois (2007). New analysis of a rat feeding study with a genetically modified maize reveals signs of hepatorenal toxicity. *Arch Environ Con Tox* **52**(4): 596-602.

Serrano, J., R. Puupponen-Pimia, A. Dauer, A. M. Aura and F. Saura-Calixto (2009). Tannins: current knowledge of food sources, intake, bioavailability and biological effects. *Mol Nutr Food Res* **53 Suppl 2**: S310-329.

Seydel, K. B., D. A. Milner, Jr., S. B. Kamiza, M. E. Molyneux and T. E. Taylor (2006). The distribution and intensity of parasite sequestration in comatose Malawian children. *J Infect Dis* **194**(2): 208-205.

Shahidi, F. and M. Naczk (1995) *Food phenolics : sources, chemistry, effects, applications*. Lancaster: Technomic Publishing Company.

Shamah-Levy, T., S. Villalpando, J. A. Rivera-Dommarco, V. Mundo-Rosas, L. Cuevas-Nasu and A. Jimenez-Aguilar (2008). Ferrous gluconate and ferrous sulfate added to a complementary food distributed by the Mexican nutrition program Oportunidades have a comparable efficacy to reduce iron deficiency in toddlers. *J Pediatr Gastroenterol Nutr* **47**(5): 660-666.

Sharieff, W., S. A. Yin, M. Wu, Q. Yang, C. Schauer, G. Tomlinson and S. Zlotkin (2006). Short-term daily or weekly administration of micronutrient Sprinkles has high compliance and does not cause iron overload in Chinese schoolchildren: a cluster-randomised trial. *Public Health Nutr* **9**(3): 336-344.

Sharieff, W., S. H. Zlotkin, W. J. Ungar, B. Feldman, M. D. Krahn and G. Tomlinson (2008). Economics of preventing premature mortality and impaired cognitive development in children through home-fortification: a health policy perspective. *Int J Technol Assess Health Care* **24**(3): 303-311.

Sharma, A. and A. C. Kapoor (1996). Levels of antinutritional factors in pearl millet as affected by processing treatments and various types of fermentation. *Plant Food Hum Nutr* **49**(3): 241-252.

Siegenberg, D., R. D. Baynes, T. H. Bothwell, B. J. Macfarlane, R. D. Lamparelli, N. G. Car, . . . F. Mayet (1991). Ascorbic-Acid Prevents the Dose-Dependent Inhibitory Effects of Polyphenols and Phytates on Nonheme-Iron Absorption. *Am J Clin Nutr* **53**(2): 537-541.

Silvie, O., M. M. Mota, K. Matuschewski and M. Prudencio (2008). Interactions of the malaria parasite and its mammalian host. *Curr Opin Microbiol* **11**(4): 352-359.

Simpore, J., F. Kabore, F. Zongo, D. Dansou, A. Bere, S. Pignatelli, . . . S. Musumeci (2006). Nutrition rehabilitation of undernourished children utilizing Spiruline and Misola. *Nutr J* **5**: 3.

Simwemba, C. G., R. C. Hoseney, E. Varrianomarston and K. Zeleznak (1984). Certain B-Vitamin and Phytic Acid Contents of Pearl-Millet [Pennisetum-Americanum (L) Leeke]. *J Agr Food Chem* **32**(1): 31-34.

Singleton, V. L., R. Orthofer and R. M. Lamuela-Raventos (1999) Analysis of total phenols and other oxidation substrates and antioxidants by means of Folin-Ciocalteu reagent. In *Oxidants and Antioxidants, Pt A.*, pp. 152-178 [L. Parker, editor]. San Diego: Academic Press.

Sisa, M., S. L. Bonnet, D. Ferreira and J. H. Van der Westhuizen (2010). Photochemistry of Flavonoids. *Molecules* **15**(8): 5196-5245.

Smeds, A. I., P. C. Eklund, R. E. Sjoholm, S. M. Willfor, S. Nishibe, T. Deyama and B. R. Holmbom (2007). Quantification of a broad spectrum of lignans in cereals, oilseeds, and nuts. *J Agr Food Chem* **55**(4): 1337-1346.

Smith, A. W., R. G. Hendrickse, C. Harrison, R. J. Hayes and B. M. Greenwood (1989). The effects on malaria of treatment of iron-deficiency anaemia with oral iron in Gambian children. *Ann Trop Paediatr Int Child Health* **9**(1): 17-23.

Snow, R. W., P. Byass, F. C. Shenton and B. M. Greenwood (1991). The relationship between anthropometric measurements and measurements of iron status and susceptibility to malaria in Gambian children. *Trans R Soc Trop Med Hyg* **85**(5): 584-589.

Snow, R. W., C. A. Guerra, A. M. Noor, H. Y. Myint and S. I. Hay (2005). The global distribution of clinical episodes of Plasmodium falciparum malaria. *Nature* **434**(7030): 214-217.

Songre-Ouattara, L. T., C. Mouquet-Rivier, C. Icard-Verniere, C. Humblot, B. Diawara and J. P. Guyot (2008). Enzyme activities of lactic acid bacteria from a pearl millet fermented gruel (ben-saalga) of functional interest in nutrition. *Int J Food Microbiol* **128**(2): 395-400.

Sosulski, F., K. Krygier and L. Hogge (1982). Free, Esterified, and Insoluble-Bound Phenolic-Acids .3. Composition of Phenolic-Acids in Cereal and Potato Flours. *J Agr Food Chem* **30**(2): 337-340.

South, P. K. and D. D. Miller (1998). Iron binding by tannic acid: effects of selected ligands. *Food Chem* **63**(2): 167-172.

Spottiswoode, N., M. Fried, H. Drakesmith and P. E. Duffy (2012). Implications of malaria on iron deficiency control strategies. *Adv Nutr* **3**(4): 570-578.

Sprinkles Global Health Initiative (2012) How are Sprinkles used. http://www.sghi.org/about_sprinkles/howare_used.html (accessed November 2012).

Sripriya, G., U. Antony and T. S. Chandra (1997). Changes in carbohydrate, free amino acids, organic acids, phytate and HCl extractability of minerals during germination and fermentation of finger millet (Eleusine coracana). *Food Chem* **58**(4): 345-350.

Steketee, R. W., B. L. Nahlen, M. E. Parise and C. Menendez (2001). The burden of malaria in pregnancy in malaria-endemic areas. *Am J Trop Med Hyg* **64**(1-2 Suppl): 28-35.

Stoltzfus, R. J. (2012). Iron and malaria interactions: programmatic ways forward. *Adv Nutr* **3**(4): 579-582.

Storcksdieck, S., G. Bonsmann and R. F. Hurrell (2007). Iron-binding properties, amino acid composition, and structure of muscle tissue peptides from in vitro digestion of different meat sources. *J Food Sci* **72**(1): S19-S29.

Suchdev, P. S., L. J. Ruth, B. A. Woodruff, C. Mbakaya, U. Mandava, R. Flores-Ayala, . . . R. Quick (2012). Selling Sprinkles micronutrient powder reduces anemia, iron deficiency, and vitamin A deficiency in young children in Western Kenya: a cluster-randomized controlled trial. *Am J Clin Nutr* **95**(5): 1223-1230.

Sun, J., J. Huang, W. X. Li, L. J. Wang, A. X. Wang, J. H. Huo, . . . C. M. Chen (2007). Effects of wheat flour fortified with different iron fortificants on iron status and anemia prevalence in iron deficient anemic students in Northern China. *Asia Pac J Clin Nutr* **16**(1): 116-121.

Sushma, D., B. K. Yadav and J. C. Tarafdar (2008). Phytate phosphorus and mineral changes during soaking, boiling and germination of legumes and pearl millet. *J Food Sci Technol* **45**(4): 344-348.

Svensson, L., B. Sekwati-Monang, D. L. Lutz, A. Schieber and M. G. Ganzle (2010). Phenolic Acids and Flavonoids in Nonfermented and Fermented Red Sorghum (Sorghum bicolor (L.) Moench). *J Agr Food Chem* **58**(16): 9214-9220.

Swain, J. H., S. M. Newman and J. R. Hunt (2003). Bioavailability of elemental iron powders to rats is less than bakery-grade ferrous sulfate and predicted by iron solubility and particle surface area. *J Nutr* **133**(11): 3546-3552.

Tanumihardjo, S. A., H. Bouis, C. Hotz, J. V. Meenakshi and B. McClafferty (2008). Biofortification of staple crops: An emerging strategy to combat hidden hunger. *Compr Rev Food Sci F* **7**(4): 329-334.

Taylor, J., J. R. N. Taylor and F. Kini (2012). Cereal biofortification: strategies, challenges, and benefits. *Cereal Food World* **57**(4): 165-169.

Taylor, J. R. and J. Dewar (2001). Developments in sorghum food technologies. *Adv Food Nutr Res* **43**: 217-264.

Teucher, B., M. Olivares and H. Cori (2004). Enhancers of iron absorption: ascorbic acid and other organic acids. *Int J Vitam Nutr Res* **74**(6): 403-419.

Thi Le, H., I. D. Brouwer, J. Burema, K. C. Nguyen and F. J. Kok (2006). Efficacy of iron fortification compared to iron supplementation among Vietnamese schoolchildren. *Nutr J* **5**: 32.

Thompson, B. A., P. A. Sharp, R. Elliott and S. J. Fairweather-Tait (2010). Inhibitory effect of calcium on non-heme iron absorption may be related to translocation of DMT-1 at the apical membrane of enterocytes. *J Agric Food Chem* **58**(14): 8414-8417.

Thompson, L. U., P. Robb, M. Serraino and F. Cheung (1991). Mammalian Lignan Production from Various Foods. *Nutr Cancer* **16**(1): 43-52.

Thuy, P. V., J. Berger, L. Davidsson, N. C. Khan, N. T. Lam, J. D. Cook, . . . H. H. Khoi (2003). Regular consumption of NaFeEDTA-fortified fish sauce improves iron status and reduces the prevalence of anemia in anemic Vietnamese women. *Am J Clin Nutr* **78**(2): 284-290.

Thuy, P. V., J. Berger, Y. Nakanishi, N. C. Khan, S. Lynch and P. Dixon (2005). The use of NaFeEDTA-fortified fish sauce is an effective tool for controlling iron deficiency in women of childbearing age in rural Vietnam. *J Nutr* **135**(11): 2596-2601.

Tidehag, P., G. Hallmans, K. Wing, R. Sjostrom, G. Agren, E. Lundin and J. X. Zhang (1996). A comparison of iron absorption from single meals and daily diets using radioFe (Fe-55,Fe-59). *Brit J Nutr* **75**(2): 281-289.

Tielsch, J. M., S. K. Khatry, R. J. Stoltzfus, J. Katz, S. C. LeClerq, R. Adhikari, . . . R. E. Black (2006). Effect of routine prophylactic supplementation with iron and folic acid on preschool child mortality in southern Nepal: community-based, cluster-randomised, placebo-controlled trial. *Lancet* **367**(9505): 144-152.

Tomas-Barberan, F. A. and M. N. Clifford (2000). Dietary hydroxybenzoic acid derivatives - nature, occurrence and dietary burden. *J Sci Food Agric* **80**(7): 1024-1032.

Tondeur, W. C., C. S. Schauer, A. L. Christofides, K. P. Asante, S. Newton, R. E. Serfass and S. H. Zlotkin (2004). Determination of iron absorption from intrinsically labeled microencapsulated ferrous fumarate (sprinkles) in infants with different iron and hematologic status by using a dual-stable-isotope method. *Am J Clin Nutr* **80**(5): 1436-1444.

Tou, E. H., J. P. Guyot, C. Mouquet-Rivier, I. Rochette, E. Counil, A. S. Traore and S. Treche (2006). Study through surveys and fermentation kinetics of the traditional processing of pearl millet (Pennisetum glaucum) into ben-saalga, a fermented gruel from Burkina Faso. *Int J Food Microbiol* **106**(1): 52-60.

Tou, E. H., C. Mouquet-Rivier, C. Picq, A. S. Traore, S. Treche and J. P. Guyot (2007). Improving the nutritional quality of ben-saalga, a traditional fermented millet-based gruel, by co-fermenting millet with groundnut and modifying the processing method. *Lwt-Food Sci Technol* **40**(9): 1561-1569.

Towo, E., E. Matuschek and U. Svanberg (2006). Fermentation and enzyme treatment of tannin sorghum gruels: effects on phenolic compounds, phytate and in vitro accessible iron. *Food Chem* **94**(3): 369-376.

Towo, E. E., U. Svanberg and G. D. Ndossi (2003). Effect of grain pre-treatment on different extractable phenolic groups in cereals and legumes commonly consumed in Tanzania. *J Sci Food Agric* **83**(9): 980-986.

Traore, T., C. Mouquet, C. Icard-Verniere, A. S. Traore and S. Treche (2004). Changes in nutrient composition, phytate and cyanide contents and alpha-amylase activity during cereal malting in small production units in Ouagadougou (Burkina Faso). *Food Chem* **88**(1): 105-114.

Tripathi, B. and K. Platel (2011). Iron fortification of finger millet (Eleucine coracana) flour with EDTA and folic acid as co-fortificants. *Food Chem* **126**(2): 537-542.

Tripathi, B., K. Platel and K. Srinivasan (2012). Double fortification of sorghum (Sorghum bicolor L. Moench) and finger millet (Eleucine coracana L. Gaertn) flours with iron and zinc. *J Cereal Sci* **55**(2): 195-201.

Tripp, K., C. G. Perrine, P. de Campos, M. Knieriemen, R. Hartz, F. Ali, . . . R. Kupka (2011). Formative research for the development of a market-based home fortification programme for young children in Niger. *Matern Child Nutr* **7 Suppl 3**: 82-95.

Troesch, B., I. Egli, C. Zeder, R. F. Hurrell, S. de Pee and M. B. Zimmermann (2009). Optimization of a phytase-containing micronutrient powder with low amounts of highly bioavailable iron for in-home fortification of complementary foods. *Am J Clin Nutr* **89**(2): 539-544.

Troesch, B., M. E. van Stujivenberg, C. M. Smuts, H. S. Kruger, R. Biebinger, R. F. Hurrell, . . . M. B. Zimmermann (2011). A micronutrient powder with low doses of highly absorbable iron and zinc reduces iron and zinc deficiency and improves weight-for-age z-scores in South african children. *J Nutr* **141**(2): 237-242.

Tuntawiroon, M., N. Sritongkul, M. Brune, L. Rossander-Hulten, R. Pleehachinda, R. Suwanik and L. Hallberg (1991). Dose-dependent inhibitory effect of phenolic compounds in foods on nonheme-iron absorption in men. *Am J Clin Nutr* **53**(2): 554-557.

Tuntawiroon, M., N. Sritongkul, L. Rossander-Hulten, R. Pleehachinda, R. Suwanik, M. Brune and L. Hallberg (1990). Rice and iron absorption in man. *Eur J Clin Nutr* **44**(7): 489-497.

Tuntipopipat, S., K. Judprasong, C. Zeder, E. Wasantwisut, P. Winichagoon, S. Charoenkiatkul, . . . T. Walczyk (2006). Chili, but not turmeric, inhibits iron absorption in young women from an iron-fortified composite meal. *J Nutr* **136**(12): 2970-2974.

Turner, G. D., V. C. Ly, T. H. Nguyen, T. H. Tran, H. P. Nguyen, D. Bethell, . . . A. R. Berendt (1998). Systemic endothelial activation occurs in both mild and severe malaria. Correlating dermal microvascular endothelial cell phenotype and soluble cell adhesion molecules with disease severity. *Am J Pathol* **152**(6): 1477-1487.

Turnlund, J. R. (1983). Use of Enriched Stable Isotopes to Determine Bioavailability of Trace-Elements in Humans. *Sci Total Environ* **28**(Jun): 385-392.

Upadhyaya, H. D., S. Ramesh, S. Sharma, S. K. Singh, S. K. Varshney, N. D. R. K. Sarma, . . . S. Singh (2011a). Genetic diversity for grain nutrients contents in a core collection of finger millet (Eleusine coracana (L.) Gaertn.) germplasm. *Field Crop Res* **121**(1): 42-52.

Upadhyaya, H. D., C. R. Ravishankar, Y. Narasimhudu, N. D. R. K. Sarma, S. K. Singh, S. K. Varshney, . . . C. L. L. Gowda (2011b). Identification of trait-specific germplasm and developing a mini core collection for efficient use of foxtail millet genetic resources in crop improvement. *Field Crop Res* **124**(3): 459-467.

Valls, J., S. Millan, M. P. Marti, E. Borras and L. Arola (2009). Advanced separation methods of food anthocyanins, isoflavones and flavanols. *J Chromatogr A* **1216**(43): 7143-7172.

van den Hombergh, J., E. Dalderop and Y. Smit (1996). Does iron therapy benefit children with severe malaria-associated anaemia? A clinical trial with 12 weeks supplementation of oral iron in young children from the Turiani Division, Tanzania. *J Trop Pediatr* **42**(4): 220-227.

van Hensbroek, M. B., S. Morris-Jones, S. Meisner, S. Jaffar, L. Bayo, R. Dackour, . . . B. M. Greenwood (1995). Iron, but not folic acid, combined with effective antimalarial therapy promotes haematological recovery in African children after acute falciparum malaria. *Trans R Soc Trop Med Hyg* **89**(6): 672-676.

van Stuijvenberg, M. E., C. M. Smuts, C. J. Lombard and M. A. Dhansay (2008). Fortifying brown bread with sodium iron EDTA, ferrous fumarate, or electrolytic iron does not affect iron status in South African schoolchildren. *J Nutr* **138**(4): 782-786.

van Stuijvenberg, M. E., C. M. Smuts, P. Wolmarans, C. J. Lombard and M. A. Dhansay (2006). The efficacy of ferrous bisglycinate and electrolytic iron as fortificants in bread in iron-deficient school children. *Brit J Nutr* **95**(3): 532-538.

Velu, G., K. N. Rai, V. Muralidharan, V. N. Kulkarni, T. Longvah and T. S. Raveendran (2007). Prospects of breeding biofortified pearl millet with high grain iron and zinc content. *Plant Breeding* **126**(2): 182-185.

Velu, G., K. N. Rai, K. L. Sahrawat and K. Sumalini (2008). Variability for grain iron and zinc content in pearl millet hybrids. *Journal of SAT agricultural research* **6**: 1-4.

Verhoef, H. (2010). Asymptomatic malaria in the etiology of iron deficiency anemia: a malariologist's viewpoint. *Am J Clin Nutr* **92**(6): 1285-1286.

Verhoef, H., C. E. West, S. M. Nzyuko, S. de Vogel, R. van der Valk, M. A. Wanga, . . . F. J. Kok (2002). Intermittent administration of iron and sulfadoxine-pyrimethamine to control anaemia in Kenyan children: a randomised controlled trial. *Lancet* **360**(9337): 908-914.

Vietmeyer, N. D. and F. R. Ruskin (1996) *Lost Crops of Africa: Grains*. Washington DC: National Academy Press.

Villalpando, S., T. Shamah, J. A. Rivera, Y. Lara and E. Monterrubio (2006). Fortifying milk with ferrous gluconate and zinc oxide in a public nutrition program reduced the prevalence of anemia in toddlers. *J Nutr* **136**(10): 2633-2637.

Virtanen, M. A., C. J. Svahn, L. U. Viinikka, N. C. Raiha, M. A. Siimes and I. E. Axelsson (2001). Iron-fortified and unfortified cow's milk: effects on iron intakes and iron status in young children. *Acta Paediatr* **90**(7): 724-731.

Viteri, F. E., E. Alvarez, R. Batres, B. Torun, O. Pineda, L. A. Mejia and J. Sylvi (1995). Fortification of Sugar with Iron Sodium Ethylenediaminotetraacetate (Fenaedta) Improves Iron Status in Semirural Guatemalan Populations. *Am J Clin Nutr* **61**(5): 1153-1163.

Viteri, F. E., R. Garciaibanez and B. Torun (1978). Sodium Iron Nafeedta as an Iron Fortification Compound in Central America - Absorption Studies. *Am J Clin Nutr* **31**(6): 961-971.

Walczyk, T., P. Kastenmayer, S. Storcksdieck Genannt Bonsmann, C. Zeder, D. Grathwohl and R. F. Hurrell (2012). Ferrous ammonium phosphate (FeNH(4)PO (4)) as a new food fortificant: iron bioavailability compared to ferrous sulfate and ferric pyrophosphate from an instant milk drink. *Eur J Nutr*.

Wall, J. S. and K. J. Carpenter (1988). Variation in Availability of Niacin in Grain Products. *Food Technol-Chicago* **42**(10): 198-&.

Walter, T., P. R. Dallman, F. Pizarro, L. Velozo, G. Pena, S. J. Bartholmey, . . . M. Arredondo (1993). Effectiveness of Iron-Fortified Infant Cereal in Prevention of Iron-Deficiency Anemia. *Pediatrics* **91**(5): 976-982.

Walter, T., P. Pino, F. Pizarro and B. Lozoff (1998). Prevention of iron-deficiency anemia: comparison of high- and low-iron formulas in term healthy infants after six months of life. *J Pediatr* **132**(4): 635-640.

Wander, K., B. Shell-Duncan and T. W. McDade (2009). Evaluation of iron deficiency as a nutritional adaptation to infectious disease: an evolutionary medicine perspective. *Am J Hum Biol* **21**(2): 172-179.

Wang, B., W. Y. Feng, T. C. Wang, J. Guang, M. Wang, J. W. Shi, . . . Z. F. Chai (2006). Acute toxicity of nano- and micro-scale zinc powder in healthy adult mice. *Toxicol Lett* **161**(2): 115-123.

Ward, H. A. and G. G. C. Kuhnle (2010). Phytoestrogen consumption and association with breast, prostate and colorectal cancer in EPIC Norfolk. *Arch Biochem Biophys* **501**(1): 170-175.

Watanabe, M. (1999). Antioxidative phenolic compounds from Japanese barnyard millet (Echinochloa utilis) grains. *J Agr Food Chem* **47**(11): 4500-4505.

Watanapaisantrakul, R., V. Chavasit and R. Kongkachuichai (2006). Fortification of soy sauce using various iron sources: Sensory acceptability and shelf stability. *Food Nutr Bull* **27**(1): 19-25.

Weaver, C. M., R. P. Heaney, B. R. Martin and M. L. Fitzsimmons (1991). Human calcium absorption from whole-wheat products. *J Nutr* **121**(11): 1769-1775.

Wegmuller, R., F. Camara, M. B. Zimmermann, P. Adou and R. F. Hurrell (2006). Salt dual-fortified with iodine and micronized ground ferric pyrophosphate affects iron status but not hemoglobin in children in Cote d'Ivoire. *J Nutr* **136**(7): 1814-1820.

Wegmuller, R., M. B. Zimmermann, D. Moretti, M. Arnold, W. Langhans and R. F. Hurrell (2004). Particle size reduction and encapsulation affect the bioavailability of ferric pyrophosphate in rats. *J Nutr* **134**(12): 3301-3304.

Welch, R. M. and R. D. Graham (2004). Breeding for micronutrients in staple food crops from a human nutrition perspective. *J Exp Bot* **55**(396): 353-364.

White, P. J. and M. R. Broadley (2009). Biofortification of crops with seven mineral elements often lacking in human diets - iron, zinc, copper, calcium, magnesium, selenium and iodine. *New Phytol* **182**(1): 49-84.

WHO (1998) *Complementary Feeding of Young Children in Developing Countries: A Review of Current Scientific Knowledge*. Geneve: World Health Organization.

WHO (2001) *Guiding Principles for Complementary Feeding of the Breastfed Child*. Geneva: World Health Organization.

WHO (2005) *Modern food biotechnology, human health and development: an evidence-based study*. Geneva: World Health Organization.

WHO (2006a) Defining and setting programme goals. In *Guidelines on food fortification with micronutrients* pp. 139-177 [L. Allen, B. De Benoist, O. D. and R. Hurrell, editors]. Geneva: World Health Organization.

WHO (2006b) Food fortification: basic principles. In *Guidelines on food fortification with micronutrients* pp. 24-37 [L. Allen, B. De Benoist, O. D. and R. Hurrell, editors]. Geneva: World Health Organization.

WHO (2006c) Iron, vitamin A and iodine. In *Guidelines on food fortification with micronutrients* pp. 97-123 [L. Allen, B. De Benoist, O. D. and R. Hurrell, editors]. Geneva: World Health Organization.

WHO (2007). Conclusions and recommendations of the WHO Consultation on prevention and control of iron deficiency in infants and young children in malaria-endemic areas. *Food Nutr Bull* **28**(4 Suppl): S621-627.

WHO (2011a) *Guideline: use of multiple micronutrient powders for home fortification of foods consumed by infants and children 6–23 months of age* Geneva: World Health Organization.

WHO (2011b) *Guideline: use of multiple micronutrient powders for home fortification of foods consumed by pregnant women* Geneva: World Health Organization.

WHO (2011c) *World malaria report 2011*. Geneva: World Health Organization.

WHO (2012). Evaluation of certain food additives: seventy-sixth report of the Joint FAO/WHO Expert Committee on Food Additives. Geneva, World Health Organization.

WHO, FAO, UNICEF, GAIN, Mi and FFI (2009) Recommendations on wheat and maize flour fortification. Meeting Report: Interim Consensus Statement. http://www.who.int/nutrition/publications/micronutrients/wheat_maize_fort.pdf

Winichagoon, P., J. E. McKenzie, V. Chavasit, T. Pongcharoen, S. Gowachirapant, A. Boonpraderm, . . . R. S. Gibson (2006). A multi micronutrient-fortif ied seasoning powder enhances the hemoglobin, zinc, and iodine status of primary school children in north east Thailand: A randomized controlled trial of efficacy, *J Nutr* **136**(6): 1617-1623.

World Health Organization (2001) *Iron deficiency anemia: assessment, prevention and control*. Geneva: WHO.

Yang, Z. Y., J. Siekmann and D. Schofield (2011). Fortifying complementary foods with NaFeEDTA - considerations for developing countries. *Matern Child Nutr* **7**: 123-128.

Yao, L. H., Y. M. Jiang, J. Shi, F. A. Tomas-Barberan, N. Datta, R. Singanusong and S. S. Chen (2004). Flavonoids in food and their health benefits. *Plant Food Hum Nutr* **59**(3): 113-122.

Yasumats.K, Nakayama.To and Chichest.Co (1965). Flavonoids of Sorghum. *J Food Sci* **30**(4): 663-&.

Zapata-Caldas, E., G. Hyman, H. Pachon, F. A. Monserrate and L. V. Varela (2009). Identifying candidate sites for crop biofortification in Latin America: case studies in Colombia, Nicaragua and Bolivia. *Int J Health Geogr* **8**.

Zimmermann, M. B. (2004). The potential of encapsulated iron compounds in food fortification: A review. *Int J Vitam Nutr Res* **74**(6): 453-461.

Zimmermann, M. B., R. Biebinger, I. Egli, C. Zeder and R. F. Hurrell (2011). Iron deficiency up-regulates iron absorption from ferrous sulphate but not ferric pyrophosphate and consequently food fortification with ferrous sulphate has relatively greater efficacy in iron-deficient individuals. *Brit J Nutr* **105**(8): 1245-1250.

Zimmermann, M. B., N. Chaouki and R. F. Hurrell (2005a). Iron deficiency due to consumption of a habitual diet low in bioavailable iron: a longitudinal cohort study in Moroccan children. *Am J Clin Nutr* **81**(1): 115-121.

Zimmermann, M. B., C. Chassard, F. Rohner, K. N'Goran E, C. Nindjin, A. Dostal, . . . R. F. Hurrell (2010). The effects of iron fortification on the gut microbiota in African children: a randomized controlled trial in Cote d'Ivoire. *Am J Clin Nutr* **92**(6): 1406-1415.

Zimmermann, M. B. and R. F. Hurrell (2007). Nutritional iron deficiency. *Lancet* **370**(9586): 511-520.

Zimmermann, M. B., R. Wegmueller, C. Zeder, N. Chaouki, R. Biebinger, R. F. Hurrell and E. Windhab (2004a). Triple fortification of salt with microcapsules of iodine, iron, and vitamin A. *Am J Clin Nutr* **80**(5): 1283-1290.

Zimmermann, M. B., R. Wegmueller, C. Zeder, N. Chaouki, F. Rohner, M. Saissi, . . . R. F. Hurrell (2004b). Dual fortification of salt with iodine and micronized ferric pyrophosphate: a randomized, double-blind, controlled trial. *Am J Clin Nutr* **80**(4): 952-959.

Zimmermann, M. B., P. Winichagoon, S. Gowachirapant, S. Y. Hess, M. Harrington, V. Chavasit, . . . R. F. Hurrell (2005b). Comparison of the efficacy of wheat-based snacks fortified with ferrous sulfate, electrolytic iron, or hydrogen-reduced elemental iron: randomized, double-blind, controlled trial in Thai women. *Am J Clin Nutr* **82**(6): 1276-1282.

Zimmermann, M. B., C. Zeder, N. Chaouki, A. Saad, T. Torresani and R. F. Hurrell (2003). Dual fortification of salt with iodine and microencapsulated iron: a randomized, double-blind, controlled trial in Moroccan schoolchildren. *Am J Clin Nutr* **77**(2): 425-432.

Zlotkin, S., P. Arthur, K. Y. Antwi and G. Yeung (2001). Treatment of anemia with microencapsulated ferrous fumarate plus ascorbic acid supplied as sprinkles to complementary (weaning) foods. *Am J Clin Nutr* **74**(6): 791-795.

Zlotkin, S., P. Arthur, C. Schauer, K. Y. Antwi, G. Yeung and A. Piekarz (2003). Home-fortification with iron and zinc sprinkles or iron sprinkles alone successfully treats anemia in infants and young children. *J Nutr* **133**(4): 1075-1080.

Zlotkin, S., S. Newton, A. Aimone, S. Amenga-Etego, K. Tchum, I. Azindow and S. Owusu-Agyei (2011). Impact of Iron Fortification on Malaria Incidence and Severity in Ghanaian Infants and Young Children. *Am J Epidemiol* **173**: S217-S217.

Zlotkin, S. H., C. Schauer, S. O. Agyei, J. Wolfson, M. C. Tondeur, K. P. Asante, . . . W. Sharieff (2006). Demonstrating zinc and iron bioavailability from intrinsically labeled microencapsulated ferrous fumarate and zinc gluconate Sprinkles in young children. *J Nutr* **136**(4): 920-925.

Zoghi, S., A. A. Mehrizi, A. Raeisi, A. A. Haghdoost, H. Turki, R. Safari, . . . S. Zakeri (2012). Survey for asymptomatic malaria cases in low transmission settings of Iran under elimination programme. *Malar J* **11**: 126.

MANUSCRIPTS

MANUSCRIPT 1 – MALARIA STUDY

Afebrile *Plasmodium falciparum* parasitemia decreases absorption of fortification iron but does not affect systemic iron utilization: a double stable-isotope study in young Beninese women

Colin I Cercamondi[1], Ines M Egli[1], Ella Ahouandjinou[2], Romain Dossa[2], Christophe Zeder[1], Lamidhi Salami[3], Harold Tjalsma[4], Erwin Wiegerinck[4], Toshihiko Tanno[5], Richard F Hurrell[1], Joseph Hounhouigan[2], Michael B Zimmermann[1,6]

[1] Laboratory for Human Nutrition, Swiss Federal Institute of Technology Zürich; [2] Laboratoire de Physiologie de la Nutrition, Université d'Abomey Calavi, Benin; [3] Médecin-coordonnateur de la zone sanitaire Atacora-Donga, Benin; [4] Department of Laboratory Medicine (Hepcidinanalysis.com), Radboud University Nijmegen Medical Centre, The Netherlands; [5] Molecular Medicine Branch, National Institutes of Health, USA; [6] Division of Human Nutrition, Wageningen University, The Netherlands

American Journal of Clinical Nutrition 2010; 92: 1385–92

Supported by:
1. INSTAPA project. The INSTAPA project receives funding from the European Community's Seventh Framework Programme [FP7/2007–2013] under grant agreement no 211484
2. Intramural Research Program of the Molecular Medicine Branch, National Institutes of Health, USA

Abstract

Background: Iron deficiency anemia (IDA) affects many young women in sub-Saharan Africa. Its etiology is multifactorial, but the major cause is low dietary iron bioavailability exacerbated by parasitic infections such as malaria.

Objectives: We investigated whether asymptomatic *Plasmodium falciparum* parasitemia in Beninese women would impair absorption of dietary iron or utilization of circulating iron.

Design: Iron absorption and utilization from an iron-fortified sorghum-based meal were estimated by using oral and intravenous isotope labels in 23 afebrile women with a positive malaria smear (asexual *P. falciparum* parasitemia; >500 parasites/μL blood). The women were studied while infected, treated, and then restudied 10 d after treatment. Iron status, hepcidin and inflammation indexes were measured before and after treatment.

Results: Treatment reduced low-grade inflammation, as reflected by decreases in serum ferritin, C-reactive protein, interleukin-6, interleukin-8, and interleukin-10 ($P < 0.05$); this was accompanied by a reduction in median serum hepcidin of ≈50%, from 2.7 to 1.4 nmol/L ($P < 0.005$). Treatment decreased serum erythropoietin and growth differentiation factor 15 ($P < 0.05$). Clearance of parasitemia increased geometric mean dietary iron absorption (from 10.2% to 17.6%; $P = 0.008$) but did not affect systemic iron utilization (85.0% compared with 83.1%; N.S.).

Conclusions: Dietary iron absorption is reduced by ≈40% in asymptomatic *P. falciparum* parasitemia, likely because of low-grade inflammation and its modulation of circulating hepcidin. Because asymptomatic parasitemia has a protracted course and is very common in malarial areas, this effect may contribute to IDA and blunt the efficacy of iron supplementation and fortification programs. This trial was registered at clinicaltrials.gov as NCT01108939.

Introduction

Iron deficiency anemia (IDA) affects many young women in sub-Saharan Africa and increases maternal and perinatal mortality and reduces work capacity (1). The etiology of IDA in Africa is multifactorial, but the major cause is low dietary iron bioavailability from monotonous cereal-based diets (2), which is exacerbated by chronic parasitic infections such as malaria (3). High-dose iron supplements can improve iron status in areas of endemic malaria (4). However, iron fortification of foods may be a more cost-effective, sustainable, and potentially safer strategy to improve iron intakes and reduce anemia (1, 5). Disappointingly, most trials of iron fortification in malarial-endemic areas of Africa have been ineffective (6) or have had only limited impact (7, 8). One reason for the blunted effect of iron fortification could be poor absorption due to chronic inflammation from parasite diseases, such as malaria.

In sub-Saharan Africa, 74% of the population lives in highly malarial endemic areas (9). In acute febrile malaria, the main mechanisms contributing to anemia are hemolysis, increased splenic clearance of erythrocytes, and, possibly, reduced erythropoesis (10, 11). The intense inflammation of acute malaria increases cytokine concentrations, such as interleukin (IL)-6 (12, 13), which can stimulate hepatic hepcidin production. High circulating hepcidin can reduce iron absorption from the gut and increase iron sequestration in the reticuloendothelial system (RES) (14) by blocking the iron-transporter ferroportin; the resulting hypoferremia limits the iron available for erythropoesis and contributes to anemia (15, 16). The use of stable isotopes showed that iron absorption was shown to be decreased in young children recovering from acute malaria (17). Growth differentiation factor (GDF-15) produced by the marrow erythroid compartment inhibits hepcidin expression and can increase iron absorption (18); however, the interaction between GDF-15 and hepcidin in malaria is unclear (19).

However, acute febrile episodes are much less frequent than asymptomatic *Plasmodium falciparum* parasitemia, which, in areas of perennial transmission, can affect much of the population for most of the year (20). Cross-sectional studies have reported high prevalence rates of 74-95% for asymptomatic *P. falciparum* parasitemia among young women in West Africa (21, 22). It is unknown whether iron bioavailability is reduced by low-level, afebrile parasitemia, which has a protracted course and only a limited inflammatory response (19, 23). If afebrile parasitemia impairs iron absorption, its contribution to iron deficiency and anemia in populations may be more important than acute febrile malaria, which usually affects individuals for only a few days each year (24).

Therefore, measuring the effect of asymptomatic parasitemia on host iron absorption and utilization may provide insights into the etiology of anemia in the tropics. Our hypotheses were as follows: 1) asymptomatic *P. falciparum* parasitemia in young African women would impair both absorption of dietary iron and utilization of circulating iron and 2) the mechanism of this effect would be the presence of low-grade inflammation affecting the balance of circulating hepcidin and GDF-15 concentrations.

Subjects and Methods

Subjects

The study was carried out in young Beninese women recruited in Natitingou and Toucountouna in the Atacora department in Northern Benin. This is an area of endemic seasonal or perennial malaria transmission The main transmission season is from May to November (25), during which time the percentage of subjects found with a positive blood test result for malaria is from ≈60% to 80% (26). The inclusion criteria were as follows: 1) female aged 16–40 y, 2) body weight <65 kg, 3) not pregnant (confirmed by pregnancy testing) and not breastfeeding, 4) no chronic medical illnesses, 5) no medicinal iron at the time of entry into the study, 6) a positive malaria smear (asexual *P. falciparum* parasitemia; >500 parasites/μL blood), 7) no clinical symptoms (eg, headache, malaise) and lack of fever (<37.5°C) or self-reported fever in past 7 d, 8) no infection with soil-transmitted helminths (hookworms, bilharziosis, *Ascaris lumbricoides*, Trichuris trichiura), 9) no severe anemia (hemoglobin <8.0 g/dL).

Sample size calculations indicated ≥16 women should be included for paired comparisons based on 80% power to detect a 40% significant difference in iron absorption, an SD of 8.2% for log transformed absorption data (based on previous Swiss Federal Institute of Technology Zurich studies), and a type I error rate of 5%. We anticipated that there could be a substantial drop-out rate during the intensive 6-wk study; therefore, we enrolled 23 subjects. The subjects provided informed written consent. Ethical approval for the study was given by the ethical review committee at the Ministry of Health in Benin and the Swiss Federal Institute of Technology Zurich (Zurich, Switzerland).

Iron absorption and utilization study

The women were studied while infected, then treated, and then restudied (**Figure 1**). Iron absorption and utilization was estimated by using stable-isotope techniques in which the incorporation into erythrocytes of an oral ^{57}Fe-dose and an intravenous ^{58}Fe-dose was measured 14 d after administration (27, 28). On day 1, a baseline venous blood sample was drawn after an overnight fast for determination of isotopic composition and iron status. The subjects then received a test meal (iron-fortified sorghum porridge; *see* below) labeled with ^{57}Fe as NaFeEDTA, which was fed under standardized conditions and close supervision. Each test meal contained 3 mg labeled ^{57}Fe. One hour later, 2 mL of an aqueous solution containing 100 μg ^{58}Fe as iron citrate was taken into a syringe and, via a 250-mL infusion bag leading into a 0.9% saline drip, slowly infused over 50 min (29). At completion of the infusion, 10 mL of normal saline was injected into the 250-mL bag, the bag was rinsed by rotation and inversion, and the saline was infused into the subject. This was done twice; we assumed there was no residual isotope remaining in the bags after this rinsing procedure. The rate of intravenous infusion of iron was based on the estimated 2 μg/min plasma appearance of iron normally absorbed from the gastrointestinal tract (30). No intake of food and fluids was allowed for 4 h after the test meal intake. Fourteen days later (day 15), a second venous blood sample was drawn after an overnight fast, and treatment and prophylaxis were started. To treat malaria, 4 tablets Malarone (250 mg atovaquone + 100 mg proguanil hydrochloride; GlaxoSmithKline, Middlesex, United Kingdom) were administered on 3 consecutive days. After treatment, one tablet Malarone was administered daily as prophylaxis until the study was completed. Before the women were restudied, there was a 10-d waiting period to allow infection-related inflammation to subside (Figure 1). During the 10 d, blood smears were repeated to check whether the women were free of malaria parasites. In addition, a stool smear was done to confirm the absence of soil-transmitted helminths. On day 25, a third venous blood sample was drawn after an overnight fast for determination of isotopic composition and confirmation of iron status. The subjects then received the second test meal and the second intravenous dose. Fourteen days later (day 39), a venous blood sample was drawn after an overnight fast. Because prophylaxis against bacterial and intestinal parasite infections is common in this area, the subjects received 500 mg ciprofloxacillin (Bayer, Leverkusen, Germany), every 2 d from day 1 (evening) through day 39 and 400 mg albendazole (GlaxoSmithKline) on days 0 and 21.

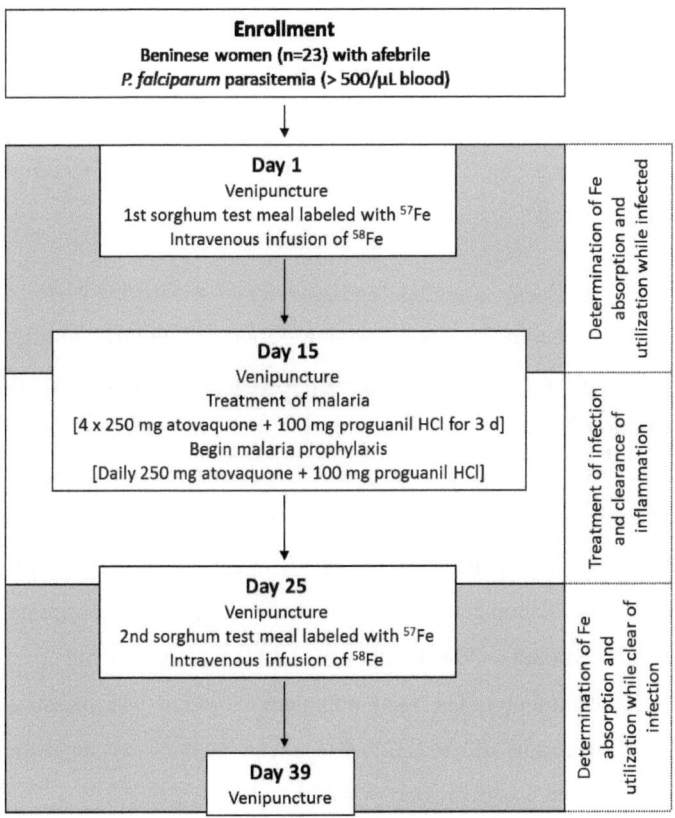

Figure 1: Overview of the study design.

Test meal preparation

A reddish sorghum variety typically found in northern Benin was soaked for 12 h and then milled. The resulting sorghum flour was fermented in excess water for 24 h, dried at 37°C in an oven, and then milled again. To prepare test meal servings, 60 g of the fermented sorghum flour was mixed with 200 g mineral water, and the resulting slurry was added to 300 g previously boiled mineral water and then boiled for 6 min. During boiling, 18 g sugar was added with continual stirring. After cooling, portions based on 50 g dry weight of sorghum flour were weighed as test meal servings. All ingredients were purchased in bulk and used for the entire study. The food portions were prepared freshly on each study day, and the 3 mg fortification iron (as $Na^{57}FeEDTA$) was added at the time of feeding after the porridge cooled.

Test meal analysis

Iron was analyzed by graphite-furnace atomic absorption spectrophotometry (GF-AAS, AA240Z; Varian, Palo Alto, CA) after a mineralization by microwave digestion (MLS ETHOSplus, MLS GmbH; Leutkirch, Germany) using a mixture of HNO_3 and H_2O_2. The phytic acid (PA) concentration was measured by using a modification of the Makower method (31), in which iron was replaced by cerium in the precipitation step. After the mineralization of food samples, inorganic phosphate was determined according to Van Veldhoven and Mannaerts (32) and converted into PA concentrations. The total polyphenol content was measured by using a modification of the Folin-Ciocalteau method, as suggested by Singleton (33) and was expressed as gallic acid equivalents.

Preparation of isotopically labeled iron

Na^{57}FeEDTA was prepared in solution from ^{57}Fe-enriched elemental iron. The metal was dissolved in 2 mL HCl, diluted with water, and stored in polytetrafluoroethylene containers. The resulting FeCl$_3$ solution was mixed with a freshly prepared aqueous Na$_2$EDTA solution (Na$_2$EDTA • H$_2$O$_2$; Sigma Chemical Co, St Louis, MO) at a molar ratio of 1:1 (Fe:EDTA) 20 min before test meal administration. The resulting NaFeEDTA solution was added to the test meal (34). Iron citrate, enriched with ^{58}Fe, was prepared for intravenous infusion from elemental ^{58}Fe according to the method described previously (35). The solution was divided in ampoules containing 100 µg Fe, sterilized, and checked for pyrogens. Enrichment of isotopic labels was 95.5% for ^{57}Fe and 93.1% for ^{58}Fe. The isotopic composition of the stable-isotope labels was measured by using negative thermal ionization-mass spectrometry (28).

Parasite diagnosis

Thick and thin blood smears were stained by using the Giemsa coloration technique and were examined independently by 2 experienced microscopists. During screening, one blood smear of the subjects was examined. After inclusion in the study, the subjects' blood smears were examined in duplicate. Parasites density was quantified against leukocytes; fields containing ≥200 leukocytes were counted; if <10 parasites were identified, the counting continued to 500 leukocytes. These counts were converted to the number of parasites per µL blood, assuming 8000 leukocytes per 1-µL blood. One hundred high-power fields were examined before a slide was declared negative. Rapid malaria diagnostic tests (Paramax-3; MD Doctors Direct GmbH, Egg b. Zurich, Switzerland) were used to support the results of the blood smears. Two KatoKatz thick smears were prepared from each stool sample. The smears were examined immediately after preparation under a light microscope, and all hookworm eggs were counted. Subsequently, the smears were re-read, and the number of eggs of *A. lumbricoides*, *Schistosoma mansoni*, and *T. trichiuria* was determined. Urine samples were collected for a pregnancy test (hCG Urine with OBC; Unipath Ltd, Bedford, United Kingdom), the detection of *Schistosoma heamatobium*, and the detection of blood, glucose, and bilirubin. *S. heamatobium* infection was assessed by using the syringe filtration technique. Ten milliliters of urine (collected between 1000 and 1400) was filtered with micropore filters, a drop of Lugol (Merck, Darmstadt, Germany) was placed on the filter, and the filter was examined under a microscope; the number of *S. haematobium* eggs was counted. Blood, glucose and bilirubin were detected by using test strips for urinalysis (Uriscan; MD Doctors Direct GmbH).

Blood analysis

Hemoglobin was measured in whole blood on the day of collection by using HemoCue hemoglobin 201+ (HemoCue AG, Wetzikon, Switzerland); anemia was defined as hemoglobin <12 g/dL (36). Serum was separated and frozen at -20 °C with no freeze-thaw cycles. Serum ferritin (SF) and high-sensitive C-reactive protein (CRP) were measured with an IMMULITE automatic system (DPC Bühlmann GmbH, Aschwil, Switzerland). Expected high-sensitivity CRP concentrations for healthy individuals were <3 mg/L. Serum transferrin receptor (TfR) was determined by using enzyme-linked immunosorbant assays (ELISA) (Ramco Laboratories Inc, Houston, TX); the normal range in healthy blood donors is 2.9–8.3 µg/mL. Iron deficiency was defined by an elevated sTfR (>8.3 µg/mL) and/or a low SF (<15 µg/L), although these indexes were likely confounded by the low-grade infection (37). IDA was defined as hemoglobin <12 g/dL and sTfR >8.3 µg/mL and/or SF <15 µg/L. Erythropoietin was measured by using ELISA (IBL ELISA; Immunobiological Laboratories, Hamburg, Germany); this assay has a reference range of 4–36 mU/mL in adults. GDF-15 was determined by ELISA for human GDF-15 (DuoSet; R&D Systems, Minneapolis, MN). Serum hepcidin measurements were performed by a combination of weak cation-exchange chromatography and time-of-flight mass spectrometry. An internal standard (synthetic hepcidin-24; Peptide International Inc, Louisville, KY) was used for quantification (38). Peptide spectra were generated on a Microflex LT matrix-enhanced laser desorption/ionisation time-of-flight mass spectrometry platform (Bruker Daltonics, Bremen, Germany). Serum hepcidin-25 concentrations were expressed as nmol/L. The lower limit of detection of this method was 0.5 nmol/L; the average CVs were 2.7% (intrarun) and 6.5% (interrun). The median reference concentration of serum hepcidin-25 in healthy Dutch adults is 4.2 nmol/L; (range: 0.5–13.9 nmol/L) (39). Total serum iron and transferrin concentrations were measured on an Aeroset (Abbott Laboratories, Abbott Park, IL) by using Roche and Abbott reagents, respectively; the transferrin concentration was converted to total iron binding capacity (TIBC) by using the equation transferrin x 25. Cytokines were analyzed in duplicate by using a Bio-Plex Pro assay (Bio-Rad Laboratories, Hercules, CA), which included: IL-2, IL-4, IL-6, IL-8, IL-10, and granulocyte macrophage colony-stimulating factor (GM-CSF), γ-interferon (IFN-γ), and tumor necrosis factor-α (TNF-α). The limit of detection for each assay was calculated as the concentration of analyte on the standard curve for which the corresponding fluorescence value was 2 SDs greater than the blank. Values below the limit of detection were reported as the limit of detection.

Whole blood was mineralized by microwave digestion, and iron was separated by anion-exchange chromatography and a subsequent solvent-solvent extraction step into diethylether. Iron was analyzed by negative thermal ionization-mass spectrometry with a magnetic sector field mass spectrometer (Finnigan MAT 262; Thermo Finnigan, Bremen, Germany) equipped with a multicollector system for simultaneous ion beam detection; isotopic dilution calculations were done as described previously (28).

Data analysis

Data were analyzed by using SPSS 13.0 for Windows (SPSS, Chicago, IL) and Excel (XP 2002; Microsoft, Seattle, WA). The amount of ^{57}Fe and ^{58}Fe label present in the blood was calculated from isotope dilution (28). Circulating iron was calculated from the blood volume based on height and weight (40) and from the hemoglobin concentration (infection period: mean of days 1 and 15; infection-free period: mean of days 25 and 39). The amount of stable isotope administered was used to calculate the fractional ^{57}Fe and ^{58}Fe incorporation into erythrocytes after 14 d. The absorption of the oral iron was calculated by dividing the percentage of erythrocyte incorporation of the oral dose by the fractional erythrocyte incorporation of the intravenous dose (30, 35). Results of iron and inflammation indexes were presented as means ± SDs if normally distributed. If not normally distributed, the results were presented as medians with 25th–75th percentile ranges. Values were logarithmically transformed before statistical analysis and tested for normality with the Shapiro-Wilk test. If normally distributed after log-transformation, paired *t* tests were used for comparisons between the 2 time points before and after treatment. If still not normally distributed after log-transformation, Wilcoxon's signed-rank tests were used for comparison. Correlations were run by using log-transformed data. For normally distributed values, Pearson's correlations were performed. Spearman's rho correlations were used for values still not normally distributed after log-transformation. Multiple linear regressions were done with log changes (day 25 to day 1) in iron absorption and utilization as the dependant variables and including the changes in log SF, log hepcidin, log TfR, log GDF-15, and log IL-10 as covariates. $P < 0.05$ was considered significant.

Results

Four hundred and thirty-six women were screened. The prevalence of *S. mansoni* and hookworm was 26% and 14%, respectively. The prevalence of asymptomatic parasitemia in women with a negative stool smear was 31%, but most of the positive smears had <500 parasites/µL blood.

The baseline characteristics of the women enrolled in the study (n = 23) are shown in **Tables 1** and **2**. The mean (± SD) age was 20.2 ± 4.8 (range: 16–35) y, and the mean (± SD) body mass index (in kg/m^2) was 21.5 ± 2.4 (range: 17.4–24.8). During the study, no subjects had detectable soil-transmitted helminths, and none developed fever or other clinical signs of infection. The median parasite value at enrollment was 1320 counts/µL. However, at entry into the study on day 1 (1–2 d later), the median parasite count had decreased to 880 parasites/µL and by day 15, 14 of the women no longer had detectable asexual *P. falciparum* parasitemia. After treatment and continuous prophylaxis, no subject had detectable parasites on days 25 and 39.

Iron indexes before and after malarial treatment are shown in Table 1. At baseline (day 1), 22% of women were anemic, the prevalence of iron deficiency was 13%, and 9% had IDA. Clearance of parasitemia was associated with a marked decrease in SF and a smaller but still significant increase in sTfR and decreases in erythropoietin and GDF-15 (Table 1 and 2). There was no significant change in serum iron or serum transferrin, whereas TIBC showed a small but significant decrease with malarial treatment (Table 1).

Infection/inflammation indexes before and after malarial treatment are shown in Table 2. Clearance of parasitemia was associated with a reduction in low-grade inflammation; there were significant decreases in CRP, IL-6, IL-8, and IL-10 between days 1 and 25. At entry into the study on day 1, 7 women had CRP concentrations >3 mg/L. After treatment, no subjects had CRP concentrations >3 mg/L. The reduction in low-grade inflammation was accompanied by a reduction in median hepcidin of ≈50%. Serum concentrations of IL-2, IL-4, GM-CSF, IFN-γ, and TNF-α (data not shown) for all subjects, except one, were below the detection limit at days 1 and 25. In the subject with detectable concentrations, concentrations on day 25 were 128.1, 1.5, 21.8, 324.2, and 40.4 pg/mL, respectively.

Table 1: Iron indexes in Beninese women ($n = 23$) before and after treatment of asymptomatic *Plasmodium falciparum* parasitemia

Iron Indexes	Day 1 (before treatment)	Day 25 (after treatment)	P
Hemoglobin (g/dL)	13.4 ± 15[1]	12.7 ± 12	0.027[2]
Serum ferritin (μg/L)	71 (29–99)[3]	37 (21–57)	0.001[4]
Serum transferrin receptor (mg/L)	5.2 (4.8–7.3)	5.6 (5.2–7.6)	0.007[2]
Total-iron-binding capacity (mmol/L)	68.0 (63.0–73.0)	63.0 (57.5–70.5)	0.006[2]
Serum iron (μmol/L)	13.3 ± 5.6	12.7 ± 6.0	0.473[2]
Transferrin saturation (%)	19.9 ± 9.1	20.7 ± 11.3	0.912[2]
Serum erythropoietin (IU/L)	11.2 (8.0–16.3)	8.9 (7.2–13.6)	0.032[2]

[1] Mean ± SD (all such values).
[2] Paired *t* test (log transformed).
[3] Median; 25th–75th percentiles in parentheses (all such values).
[4] Wilcoxon's signed-rank test (log transformed).

Table 2: Inflammation indexes in Beninese women ($n = 23$) before and after treatment of asymptomatic *Plasmodium falciparum* parasitemia[1]

Inflammation indexes	Day 1 (before treatment)	Day 25 (after treatment)	P
Parasites (no./µL)[1]	880 (123–2760)	0 (0)	0.001[2]
C-reactive protein (mg/L)	0.9 (0.4–5.7)	0.30 (0.3–0.4)	0.001[2]
Hepcidin (nmol/L)	2.7 (1.0–4.6)	1.4 (0.7–2.4)	0.005[2]
GDF-15 (pg/mL)	497 (400–612)	381 (341–432)	0.003[2]
Interleukin-6 (pg/mL)	1.32 (0.96–1.92)	1.27 (0.70–1.32)	0.031[2]
Interleukin-8 (pg/mL)	7.31 (4.58–10.03)	4.18 (2.60–5.67)	0.001[3]
Interleukin-10 (pg/mL)	7.38 (4.44–13.93)	2.94 (1.91–2.94)	0.001[3]

[1] All values are medians; 25th–75th percentiles in parentheses. GDF-15, growth differentiation factor 15.
[2] Wilcoxon's signed-rank test (log transformed).
[3] Paired *t* test (log transformed).

The native iron content of the sorghum used to prepare the test meals was 3.4 ± 0.2 mg/100 g, but the total iron content in the sorghum flour was 8.9 ± 0.5 mg/100 g due to the introduction of iron, which occurred during local milling. Phytate and polyphenol contents of the sorghum flour were 421 ± 24 mg and 37.5 ± 0.5 mg/100 g, respectively. The meal based on 50 g flour therefore contained 1.7 mg native Fe, 2.8 mg contaminant Fe, 3 mg Fe as NaFeEDTA, 211 mg phytate and 18.8 mg polyphenols.

Iron absorption and utilization before and after malarial treatment are shown in **Figures 2** and **3**. There was a significant 70% increase in oral iron absorption with treatment: geometric mean absorption (95% CI) on days 1 and 25 were 10.2% (7.4–14.0%) and 17.6% (13.5–22.3%), respectively; 18 of 23 women showed an increase in absorption (Figure 2). In contrast, there was no significant change in systemic utilization of the intravenous iron; 14 of 23 women showed a decrease in utilization after treatment (Figure 3).

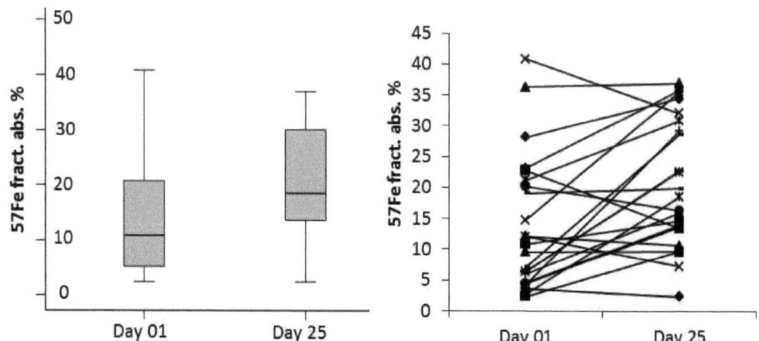

Figure 2: Iron absorption in Beninese women ($n = 23$) who consumed a sorghum porridge labeled with 3 mg ^{57}Fe as NaFeEDTA before (day 1) and after (day 25) malarial treatment. Iron absorption on day 1 was significantly lower than that on day 25 (Wilcoxon's signed-rank test, $P = 0.008$). The box plots show the median and 25th and 75th percentiles. Whiskers in the plots represent the highest and lowest values. The line graph shows individual iron absorption of the 23 women before and after treatment. fract. abs., fractional absorption.

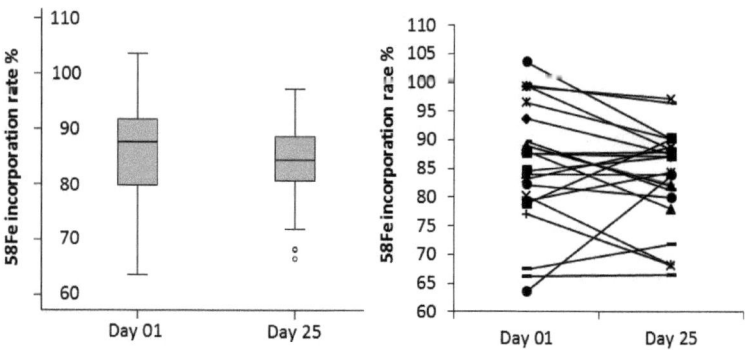

Figure 3: Erythrocyte iron utilization in Beninese women ($n = 23$) who received an intravenous dose of 100 µg ^{58}Fe labeled iron citrate before (day 1) and after (day 25) malarial treatment. Iron utilization did not differ significantly between the 2 d (Wilcoxon's signed-rank test, $P = 0.107$). The box plots show the median and 25th and 75th percentiles. Whiskers in the plots represent the highest and lowest values that are not outliers. Outliers (values that are between 1.5 and 3 times the interquartile range) are represented by circles beyond the whiskers. The line graph shows individual iron utilization of the 23 women before and after treatment.

Associations between iron absorption and iron and inflammation indexes are shown in **Table 3**. On day 1, during infection, SF, CRP and GDF-15 showed significant negative correlations with iron absorption, whereas TfR and TIBC showed significant positive correlations. After malarial treatment (day 25), SF, serum iron, serum transferrin, and hepcidin showed significant negative correlations with iron absorption. On day 1, systemic iron utilization negatively correlated with GDF-15 ($r = -0.578$, $P < 0.01$); otherwise, there were no relevant significant associations between the measured indexes and systemic utilization. Change (Δ) in iron absorption (from days 25 to day 1) was negatively correlated with ΔSF ($r = -0.493$, $P < 0.05$), ΔCRP ($r = -0.574$, $P < 0.01$), and ΔIL-8 ($r = -0.426$, $P = 0.05$) and was positively correlated with ΔTIBC ($r = 0.415$, $P < 0.05$). ΔIron utilization (day 25 – day 1) was positively correlated with ΔIL-10 ($r = 0.422, P < 0.05$).

Table 3: Univariate correlations of iron absorption with iron and inflammation indexes on day 1 (before malarial treatment) and day 25 (after malarial treatment) in Beninese women ($n = 23$) with asymptomatic *Plasmodium falciparum* parasitemia[1]

	Iron absorption on day 1	Iron absorption on day 25
Serum ferritin[2]	–0.554[3]	–0.784[3]
Serum transferrin receptor[4]	0.576[3]	—
Total-iron-binding capacity[4]	0.472[5]	—
Serum iron[2]	—	–0.473[5]
Transferrin saturation[2]	—	–0.456[5]
C-reactive protein[2]	–0.491[5]	—
Hepcidin[2]	—	–0.679[3]
GDF-15[4]	–0.446[5]	—

[1] GDF-15, growth differentiation factor 15.
[2] Spearman's rho correlations.
[3] $P < 0.01$ (2-tailed).
[4] Pearson's correlation coefficients.
[5] $P < 0.05$ level (2-tailed).

In multivariate models with log Δiron absorption and Δutilization as the dependent variables and including log ΔSF, Δhepcidin, ΔTfR, ΔGDF-15, and ΔIL-10 as covariates, ΔSF ($P = 0.008$), ΔGDF-15 ($P = 0.017$), and ΔTfR ($P = 0.025$) were significant predictors of Δiron absorption ($r^2 = 0.598$), and ΔGDF-15 ($P < 0.000$) and ΔIL-10 ($P < 0.001$) were predictors of Δiron utilization ($r^2 = 0.621$).

Discussion

This study showed that afebrile malarial parasitemia, a common condition in endemic areas, decreases dietary iron absorption. This effect appears to be due to low-level inflammation modulation of serum hepcidin. This effect may help explain why iron supplements and iron fortification of staple foods or complementary foods may be less effective in malarial endemic areas (6–8). Fractional iron absorption in the present study was relatively high, which was likely due to the use of NaFeEDTA, an iron chelator, as the iron fortificant (41–45). The test meal in the present study included ≈2.8 mg contaminant Fe introduced by local milling; we assumed its bioavailability in our test meal was negligible (46, 47).

Our study is the first to directly quantify iron incorporation into erythrocytes from dietary iron and systemic iron during, and after treatment, of asymptomatic malarial parasitemia. The advantage of using both an oral and an intravenous tracer is that it allows separation of dietary iron bioavailability into its 2 components: intestinal absorption and systemic utilization. Acute febrile malaria causes sequestration of RES iron and hypoferremia (48), which often coexists with normal or increased bone marrow iron (49). Release of iron from macrophages, like intestinal cells, is impaired with increased hepcidin and responsible for anemia by restricting iron supply to erythrocytes. In contrast with subjects with acute afebrile malaria, our subjects with afebrile parasitemia had no evidence of hypoferremia during infection: the decrease in hepcidin with parasite clearance was not associated with a change in serum iron or in the percentage of serum transferrin. In contrast, asymptomatic parasitemia decreases dietary iron absorption, which suggests that iron export via ferroportin from enterocytes is sensitive to even small changes in hepcidin in the low-normal range. This finding is consistent with a previous cross-sectional study that used stable isotopes in healthy Swiss women, in which mean iron absorption from a test meal increased and serum hepcidin concentrations decreased slightly (50).

Previous studies of iron metabolism and hepcidin during malaria have focused on changes during treatment of acute febrile infection. Hepcidin regulation is complex during malaria infection: hepcidin production may be upregulated by inflammatory cytokines (15, 51, 52); however, at the same time, hypoxia associated with malarial anemia and possibly with erythropoetic factors such as GDF-15, may down-regulate hepcidin production (53, 54). The inflammatory stimuli appear to predominate, because studies have generally found a positive relation between malaria and circulating hepcidin (55). In anemic Tanzanian children <3 y of age with febrile *P. falciparum* malaria, urinary hepcidin strongly increased and was positively correlated with inflammation; antimalarial treatment led to a rapid decrease in urinary hepcidin, TNF-α, IL-6, and IL-10 and in a reversal of hypoferremia (48).

In contrast with previous studies, our Beninese subjects had lower levels of parasitemia, no clinical symptoms and probable partial immunity from repeated exposure (56). Treatment of parasitemia decreased inflammation (CRP, SF, IL-6, IL-8, and IL-10), which was associated with a significant fall in serum hepcidin. Thus, our findings suggest that the low-grade inflammation in asymptomatic malaria is sufficient to trigger a small increase in serum hepcidin. Similarly, because SF is an acute phase protein, the resolution of low-grade inflammation likely explains the ≈48% decrease in SF associated with treatment of parasitemia (Table 1). Iron absorption was lower in those subjects with more inflammation (as reflected in higher SF and/or higher CRP concentrations) on day 1 or higher hepcidin on day 25 (Table 3). During experimental *P. falciparum* malaria infection in Dutch adults (57), low parasitemia and its treatment was associated with an increase in IL-6, a decline in serum iron and serum transferrin, and an increase in SF; there was an increase in serum hepcidin in 4 of 5 volunteers. Parasitemia and its treatment was associated with a transient decrease in reticulocyte hemoglobin content over 3 d, which was followed by a compensatory reticulocytosis, which suggests impaired iron incorporation into developing erythrocytes. In contrast, we directly measured systemic iron utilization by using labeled intravenous iron and found no effect of treatment of parasitemia. This suggests that parasitemia-associated inflammation may reduce iron release from macrophages rather than impair iron incorporation during erythropoesis. Compared with previous studies, geometric mean incorporation of an intravenous iron dose in the present study (83–85%) was lower than in young Thai women (93%) (29), but was somewhat higher than in older British men (79%) (30).

In our study, there were small but significant decreases in both erythropoietin and GDF-15 with clearance of parasitemia. Acute malaria may impair iron homeostasis through a direct and/or indirect suppressive effect on erythropoesis (11, 58). Increased hepcidin production may contribute by directly inhibiting erythroid progenitor proliferation and survival (59). Decreased erythropoietin production and/or sensitivity during malaria may also play a role (10, 60, 61), but most studies have found an adequate host erythropoietin response to malaria infection (62). In the study by Doherty et al (17), erythropoietin was significantly lower in children recovering from febrile malaria than in anemic children without malaria. In contrast, in our study of mostly nonanemic women, there was a decrease in erythropoietin with treatment, which suggested that erythropoietin was not limiting erythropoesis during asymptomatic parasitemia. A small but significant fall in serum GDF-15 with clearance of parasitemia was observed. Because circulating GDF-15 fell along with hepcidin, the improvement in iron absorption did not appear to be the result of a "marrow" signal from GDF-15 to increase iron absorption, but rather to a decrease in circulating hepcidin.

In our studies, we did not have an additional control group of women without malaria who received antimalarial treatment and then underwent measurements of iron absorption and/or utilization. Thus, we cannot entirely rule out a potential effect of the antimalarials agents per se on iron metabolism, independent of their effect on *P. falciparum*. We administered prophylactic ciprofloxacillin and albendazole during the 6-wk study period to avoid potential confounding of the results by common local infections and associated inflammation. This was successful in that none of the subjects developed an infection during the study that would have led to their exclusion. Although ciprofloxacillin showed some in vitro activity against the blood stage of *Plasmodium* spp. (63, 64), it was not effective against *Plasmodium* infections (65, 66). Therefore, the negative blood smears in several of the subjects on day 15 (before antimalarial treatment) were likely explained either by self-clearance, which is thought to commonly occur in endemic areas (67, 68), or by the lack of sensitivity of blood smears to detect subpatent (eg, liver stage) parasitemia (69).

Afebrile parasitemia is very common in endemic areas and often has a protracted course (19, 23). Because it reduces iron absorption, it may be an important contributor to IDA in young women with poor diets but high iron requirements because of menstrual blood losses and repeated pregnancies. In this context, the benefits of intermittent treatment of malaria on anemia rates (70, 71) may be due not only to a reduction in acute malarial anemia, but also to the periodic clearance of asymptomatic parasitemia and a resulting improvement in dietary iron absorption.

Acknowledgments

We thank the subjects for their participation in the study and the nursing and laboratory staff at the Hopital du Zone Natitingou (Natitingou, Benin). We also thank Traore Mahamadou (Centre Suisse de Recherches Scientifiques, Cote d'Ivoire) and Siem Klaver (Radboud University Nijmegen Medical Centre, Nijmegen, Netherlands) for laboratory assistance.

The authors' responsibilities were as follows—CIC, IME, CZ, RD, JH, LS, RFH, and MBZ. designed the research; CIC, EA, IME, and CZ: conducted the research; CIC, CZ, TT, HT, and EW: analyzed the data; and CIC, IME, RFH, and MBZ: wrote the manuscript and had primary responsibility for the final content of the manuscript. All authors read and approved the final version of the manuscript. None of the authors had a conflict of interest with regard to this manuscript.

References

1. Zimmermann MB, Hurrell RF. Nutritional iron deficiency. Lancet 2007;370:511-520.
2. Zimmermann MB, Chaouki N, Hurrell RF. Iron deficiency due to consumption of a habitual diet low in bioavailable iron: a longitudinal cohort study in Moroccan children. American Journal of Clinical Nutrition 2005;81:115-121.

3. Newton CRJC, Warn PA, Winstanley PA, et al. Severe anaemia in children living in a malaria endemic area of Kenya. Tropical Medicine & International Health 1997;2:165-178.
4. Menendez C, Schellenberg D, Quinto L, et al. The effects of short-term iron supplementation on iron status in infants in malaria-endemic areas. American Journal of Tropical Medicine and Hygiene 2004;71:434-440.
5. WHO, FAO. Guidelines on food fortification with micronutrients Geneva: WHO/FAO, 2006.
6. Rohner F, Zimmermann MB, Amon RJ, et al. In a Randomized Controlled Trial of Iron Fortification, Anthelmintic Treatment and Intermittent Preventive Treatment of Malaria for Anemia Control in Ivorian Children, only Anthelmintic Treatment Shows Modest Benefit. J. Nutr. 2010:jn.109.114256.
7. Wegmuller R, Camara F, Zimmermann MB, Adou P, Hurrell RF. Salt dual-fortified with iodine and micronized ground ferric pyrophosphate affects iron status but not hemoglobin in children in Cote d'Ivoire. Journal of Nutrition 2006;136:1814-1820.
8. Andang'o PEA, Osendarp SJM, Ayah R, et al. Efficacy of iron-fortified whole maize flour on iron status of schoolchildren in Kenya: a randomised controlled trial. Lancet 2007;369:1799-1806.
9. WHO. The World Health Report 1999: Making a Difference. Geneva: World Health Organization, 1999.
10. Jelkmann W. Proinflammatory cytokines lowering erythropoietin production. Journal of Interferon and Cytokine Research 1998;18:555-559.
11. Lamikanra AA, Brown D, Potocnik A, Casals-Pascual C, Langhorne J, Roberts DJ. Malarial anemia: of mice and men. Blood 2007;110:18-28.
12. Wenisch C, Linnau KF, Looaresuwan S, Rumpold H. Plasma levels of the interleukin-6 cytokine family in persons with severe Plasmodium falciparum malaria. Journal of Infectious Diseases 1999;179:747-750.
13. Lyke KE, Burges R, Cissoko Y, et al. Serum Levels of the Proinflammatory Cytokines Interleukin-1 Beta (IL-1(72)), IL-6, IL-8, IL-10, Tumor Necrosis Factor Alpha, and IL-12(p70) in Malian Children with Severe Plasmodium falciparum Malaria and Matched Uncomplicated Malaria or Healthy Controls. Infect. Immun. 2004;72:5630-5637.
14. Weiss G. Iron metabolism in the anemia of chronic disease. Biochimica Et Biophysica Acta-General Subjects 2009;1790:682-693.
15. Nemeth E, Rivera S, Gabayan V, et al. IL-6 mediates hypoferremia of inflammation by inducing the synthesis of the iron regulatory hormone hepcidin. Journal of Clinical Investigation 2004;113:1271-1276.
16. Lee P, Peng HF, Gelbart T, Wang L, Beutler E. Regulation of hepcidin transcription by interleukin-1 and interleukin-6. Proceedings of the National Academy of Sciences of the United States of America 2005;102:1906-1910.
17. Doherty CP, Cox SE, Fulford AJ, et al. Iron Incorporation and Post-Malaria Anaemia. Plos One 2008;3:-.
18. Tanno T, Bhanu NV, Oneal PA, et al. High levels of GDF15 in thalassemia suppress expression of the iron regulatory protein hepcidin. Nature Medicine 2007;13:1096-1101.
19. de Mast Q, Syafruddin D, Keijmel S, et al. Increased serum hepcidin and alterations in blood iron parameters associated with asymptomatic P. falciparum and P. vivax malaria. Haematologica 2010:haematol.2009.019331.
20. Eke RA, Chigbu LN, W. N. High Prevalence of Asymptomatic Plasmodium Infection in a Suburb of Aba Town, Nigeria. Annals of African Medicine 2006;5:42-45.
21. Agan TU, Ekabua JE, Iklaki CU, Oyo-Ita A, Ibanga I. Prevalence of asymptomatic malaria parasitaemia. Asian Pacific Journal of Tropical Medicine 2010;3:51-55.
22. Onyenekwe CC, Meludu SC, Dioka CE, Salimonu LS. Prevalence of asymptomatic malaria parasitaemia amongst pregnant women. Indian journal of malariology 2002;39:60-65.
23. Imrie H, Fowkes FJI, Michon P, Tavul L, Reeder JC, Day KP. Low prevalence of an acute phase response in asymptomatic children from a malaria-endemic area of Papua New Guinea. American Journal of Tropical Medicine and Hygiene 2007;76:280-284.
24. Carneiro IA, Smith T, Lusingu JPA, Malima R, Utzinger J, Drakeley CJ. Modeling the Relationship between the Population Prevalence of *Plasmodium Falciparum* Malaria and Anemia. American Journal of Tropical Medicine and Hygiene 2006;75:82-89.
25. MARA/ARMA Collaboration. Malaria Seasonality Model. www.mara.org.za, 2001.
26. MARA/ARMA Collaboration. Malaria Prevalence Model. www.mara.org.za, 2002.

27. Kastenmayer P, Davidsson L, Galan P, Cherouvrier F, Hercberg S, Hurrell RF. A Double Stable-Isotope Technique for Measuring Iron-Absorption in Infants. British Journal of Nutrition 1994;71:411-424.
28. Walczyk T, Davidsson L, Zavaleta N, Hurrell RF. Stable isotope labels as a tool to determine the iron absorption by Peruvian school children from a breakfast meal. Fresenius Journal of Analytical Chemistry 1997;359:445-449.
29. Zimmermann MB, Fucharoen S, Winichagoon P, et al. Iron metabolism in heterozygotes for hemoglobin E (HbE), alpha-thalassemia 1, or beta-thalassemia and in compound heterozygotes for HbE/beta-thalassemia. American Journal of Clinical Nutrition 2008;88:1026-1031.
30. Roe MA, Heath ALM, Oyston SL, et al. Iron absorption in male C282Y heterozygotes. American Journal of Clinical Nutrition 2005;81:814-821.
31. Makower RU. Extraction and Determination of Phytic Acid in Beans (Phaseolus-Vulgaris). Cereal Chem 1970;47:288-96.
32. Vanveldhoven PP, Mannaerts GP. Inorganic and Organic Phosphate Measurements in the Nanomolar Range. Anal Biochem 1987;161:45-48.
33. Singleton VL. Citation Classic - Colorimetry of Total Phenolics with Phosphomolybdic-Phosphotungstic Acid Reagents. Current Contents/Agriculture Biology & Environmental Sciences 1985:18-18.
34. Davidsson L, Ziegler E, Zeder C, Walczyk T, Hurrell R. Sodium iron EDTA [NaFe(III)EDTA] as a food fortificant: erythrocyte incorporation of iron and apparent absorption of zinc, copper, calcium, and magnesium from a complementary food based on wheat and soy in healthy infants. American Journal of Clinical Nutrition 2005;81:104-109.
35. Dainty JR, Roe MA, Teucher B, Eagles J, Fairweather-Tait SJ. Quantification of unlabelled non-haem iron absorption in human subjects: a pilot study. British Journal of Nutrition 2003;90:503-506.
36. WHO/UNICEF/UNU, ed. Iron Deficiency Anemia Assessment, Prevention and Control. Geneve: World Health Organization, 2001.
37. Zimmermann MB. Methods to assess iron and iodine status. British Journal of Nutrition 2008;99:S2-S9.
38. Swinkels DW, Girelli D, Laarakkers C, et al. Advances in Quantitative Hepcidin Measurements by Time-of-Flight Mass Spectrometry. Plos One 2008;3:-.
39. Kroot JJC, Hendriks JCM, Laarakkers CMM, et al. (Pre)analytical imprecision, between-subject variability, and daily variations in serum and urine hepcidin: Implications for clinical studies. Analytical Biochemistry 2009;389:124-129.
40. Geigy Scientific Tables. Hematology and human genetics. Basel, Switzerland: Ciba-Geigy Limited, 1979.
41. Ballot DE, Macphail AP, Bothwell TH, Gillooly M, Mayet FG. Fortification of Curry Powder with NaFe(III)EDTA in an Iron-Deficient Population - Report of a Controlled Iron-Fortification Trial. American Journal of Clinical Nutrition 1989;49:162-169.
42. Garby L, Areekul S. Iron Supplementation in Thai Fish-Sauce. Annals of Tropical Medicine and Parasitology 1974;68:467-476.
43. Davidsson L, Dimitriou T, Boy E, Walczyk T, Hurrell RF. Iron bioavailability from iron-fortified Guatemalan meals based on corn tortillas and black bean paste. American Journal of Clinical Nutrition 2002;75:535-539.
44. Macphail AP, Bothwell TH, Torrance JD, et al. Factors Affecting the Absorption of Iron from Fe(III)EDTA. British Journal of Nutrition 1981;45:215-227.
45. Layrisse M, Martineztorres C. Fe(III)-EDTA Complex as Iron Fortification. American Journal of Clinical Nutrition 1977;30:1166-1174.
46. Derman DP, Bothwell TH, Torrance JD, et al. Iron-Absorption from Ferritin and Ferric Hydroxide. Scandinavian Journal of Haematology 1982;29:18-24.
47. Derman DP, Bothwell TH, Torrance JD, et al. Iron-Absorption from Maize (Zea-Mays) and Sorghum (Sorghum-Vulgare) Beer. British Journal of Nutrition 1980;43:271-279.
48. de Mast Q, Nadjm B, Reyburn H, et al. Assessment of Urinary Concentrations of Hepcidin Provides Novel Insight into Disturbances in Iron Homeostasis during Malarial Infection. Journal of Infectious Diseases 2009;199:253-262.
49. Phillips RE, Looareesuwan S, Warrell DA, et al. The Importance of Anemia in Cerebral and Uncomplicated Falciparum-Malaria - Role of Complications, Dyserythropoiesis and Iron Sequestration. Quarterly Journal of Medicine 1986;58:305-323.

50. Zimmermann MB, Troesch B, Biebinger R, Egli I, Zeder C, Hurrell RF. Plasma hepcidin is a modest predictor of dietary iron bioavailability in humans, whereas oral iron loading, measured by stable-isotope appearance curves, increases plasma hepcidin. American Journal of Clinical Nutrition 2009;90:1280-1287.
51. Kemna E, Pickkers P, Nemeth E, van der Hoeven H, Swinkels D. Time-course analysis of hepcidin, serum iron, and plasma cytokine levels in humans injected with LPS. Blood 2005;106:1864-1866.
52. Wrighting DM, Andrews NC. Interleukin induces hepcidin expression through STAT3. Blood 2006;108:3204-3209.
53. Pak M, Lopez MA, Gabayan V, Ganz T, Rivera S. Suppression of hepcidin during anemia requires erythropoietic activity. Blood 2006;108:3730-3735.
54. Peyssonnaux C, Zinkernagel AS, Schuepbach RA, et al. Regulation of iron homeostasis by the hypoxia-inducible transcription factors (HIFs). Journal of Clinical Investigation 2007;117:1926-1932.
55. Howard CT, McKakpo US, Quakyi IA, et al. Relationship of hepcidin with parasitemia and anemia among patients with uncomplicated Plasmodium falciparum malaria in Ghana. American Journal of Tropical Medicine and Hygiene 2007;77:623-626.
56. Hviid L. Naturally acquired immunity to Plasmodium falciparum malaria in Africa. Acta Tropica 2005;95:270-275.
57. de Mast Q, van Dongen-Lases EC, Swinkels DW, et al. Mild increases in serum hepcidin and interleukin-6 concentrations impair iron incorporation in haemoglobin during an experimental human malaria infection. British Journal of Haematology 2009;145:657-664.
58. Wang CQ, Udupa KB, Lipschitz DA. Interferon-Gamma Exerts Its Negative Regulatory Effect Primarily on the Earliest Stages of Murine Erythroid Progenitor-Cell Development. Journal of Cellular Physiology 1995;162:134-138.
59. Dallalio G, Law E, Means RT. Hepcidin inhibits in vitro erythroid colony formation at reduced erythropoietin concentrations. Blood 2006;107:2702-2704.
60. Spivak JL. The blood in systemic disorders. Lancet 2000;355:1707-1712.
61. Wickramasinghe SN, Abdalla SH. Blood and bone marrow changes in malaria. Best Practice & Research Clinical Haematology 2000;13:277-299.
62. Nweneka CV, Doherty CP, Cox S, Prentice A. Iron delocalisation in the pathogenesis of malarial anaemia. Transactions of the Royal Society of Tropical Medicine and Hygiene 2009;104:175-184.
63. Mahmoudi N, Ciceron L, Franetich JF, et al. In vitro activities of 25 quinolones and fluoroquinolones against liver and blood stage Plasmodium spp. Antimicrobial Agents and Chemotherapy 2003;47:2636-2639.
64. Goodman CD, Su V, McFadden GI. The effects of anti-bacterials on the malaria parasite Plasmodium falciparum. Molecular and Biochemical Parasitology 2007;152:181-191.
65. Watt G, Shanks GD, Edstein MD, Pavanand K, Webster HK, Wechgritaya S. Ciprofloxacin Treatment of Drug-Resistant Falciparum-Malaria. Journal of Infectious Diseases 1991;164:602-604.
66. Stromberg A, Bjorkman A. Ciprofloxacin Does Not Achieve Radical Cure of Plasmodium-Falciparum Infection in Sierra-Leone. Transactions of the Royal Society of Tropical Medicine and Hygiene 1992;86:373-373.
67. Takala S, Branch O, Escalante AA, Kariuki S, Wootton J, Lal AA. Evidence for intragenic recombination in Plasmodium falciparum: identification of a novel allele family in block 2 of merozoite surface protein-1: Asembo Bay Area Cohort Project XIV. Molecular and Biochemical Parasitology 2002;125:163-171.
68. Patarroyo ME, Amador R, Clavijo P, et al. A synthetic vaccine protects humans against challenge with asexual blood stages of Plasmodium falciparum malaria. Nature 1988;332:158-161.
69. Bottius E, Guanzirolli A, Trape JF, Rogier C, Konate L, Druilhe P. Malaria: Even more chronic in nature than previously thought; Evidence for subpatent parasitaemia detectable by the polymerase chain reaction. Transactions of the Royal Society of Tropical Medicine and Hygiene 1996;90:15-19.
70. Greenwood B. Review: Intermittent preventive treatment - a new approach to the prevention of malaria in children in areas with seasonal malaria transmission. Tropical Medicine & International Health 2006;11:983-991.

71. Odhiambo FO, Hamel MJ, Williamson J, et al. Intermittent Preventive Treatment in Infants for the Prevention of Malaria in Rural Western Kenya: A Randomized, Double-Blind Placebo-Controlled Trial. Plos One 2010;5: e10016.

Manuscript 2 – Sorghum Polyphenol Study

Sodium iron EDTA and ascorbic acid, but not polyphenol oxidase treatment, counteract the strong inhibitory effect of polyphenols from brown sorghum on the absorption of fortification iron in young women

Colin I Cercamondi[1], Ines M Egli[1], Christophe Zeder[1] and Richard F Hurrell[1]

[1] Laboratory of Human Nutrition, Institute of Food, Nutrition and Health, ETH Zurich, Zurich, Switzerland

British Journal of Nutrition, Cambridge University Press, available on CJO 19th August 2013, doi: 10.1017/S0007114513002705

Supported by the INSTAPA project, which receives funding from the European Community's Seventh Framework Programme [FP7/2007–2013] under grant agreement no. 211484

Abstract

In addition to phytate, polyphenols (PPs) might contribute to low iron bioavailability from sorghum-based foods. To investigate the inhibiting effects of sorghum PPs on iron absorption and the potential enhancing effect of ascorbic acid (AA), NaFeEDTA and the polyphenol oxidase enzyme laccase, we performed three iron absorption studies in 50 young women consuming dephytinized iron-fortified test meals based on white and brown sorghum varieties with different PPs concentrations. Iron absorption was measured as erythrocyte incorporation of stable iron isotopes. In study 1, iron absorption from meals with 17 mg PPs (8·5%) was higher than from meals with 73 mg PPs (3·2%) and 167 mg PPs (2·7%; $P<0·001$). Absorption from meals containing 73 and 167 mg PPs did not differ ($P=0·9$). In study 2 iron absorption from NaFeEDTA-fortified meals (167 mg PPs) was higher than from the same meals fortified with $FeSO_4$ (4·6% vs. 2·7%; $P<0·001$) but still lower than from $FeSO_4$-fortified meals with 17 mg PPs (10·7%; $P<0·001$). In study 3, laccase treatment decreased PPs from 167 to 42 mg (4·8%) but did not improve absorption compared with meals with 167 mg PPs (4·6%; $P=0·4$), whereas adding AA increased absorption to 13·6% ($P<0·001$). These findings suggest that PPs from brown sorghum contribute to low iron bioavailability from sorghum foods and that AA and to lesser extent NaFeEDTA, but not laccase, have the potential to overcome PP inhibition and improve iron absorption from sorghum foods.

Introduction

Iron deficiency (ID) is the most prevalent micronutrient deficiency worldwide, affecting principally children <5 years of age and women of childbearing age living in the poorer communities of the developing world [1]. ID has major negative impacts on health and mental development, and in pregnancy contributes to the risk of severe anemia, which is associated with higher maternal morbidity and mortality [2, 3]. The etiology of ID in developing countries is multifactorial, but major contributors are low dietary iron bioavailability and intake from monotonous diets based on cereal staple food such as sorghum [4].

Sorghum [*Sorghum bicolor* (L.) Moench] is the world's fifth most important cereal in terms of production and a staple food in sub-Saharan Africa and India [5] where it is a major source of calories, proteins and micronutrients for millions of people [6]. The color of sorghum grain varies among different varieties. White grains are preferably used for foods such as porridge, baked goods and couscous. Red and brown grains are also widely used for porridges and/or pastes but in addition are used for the preparation of fermented and non-fermented local beverages [7-9]. Thin sorghum porridges, which are occasionally fermented, are popular complementary foods throughout sub-Saharan Africa [10-13].

The native iron concentration of sorghum has been reported to be about 3–5 mg/100 g flour [14]. Iron bioavailability from sorghum-based diets is expected to be low due to high amounts of phytic acid (PA) in the grains [15, 16] and high amounts of polyphenols (PPs) in some colored varieties [7, 17]. Both, PA and PPs, can be potent iron absorption inhibitors, forming non-absorbable complexes in the gut lumen [18, 19]. The inhibitory effect of PPs varies according to their structure [19-21]. The total PP concentration in sorghum depends on genetic factors such as seed color and thickness of the pericarp, but also on environmental factors such as growing conditions [22]. Sorghum varieties with red and brown colors have higher average PP concentrations (1·35%) than light colored sorghum grains (0·40%) [7]; the color is mainly attributable to PPs called deoxyanthocyanidins [23, 24].

Fortification of cereal flours with iron is the main strategy to combat ID in communities where cereals are the staple food. Unlike wheat, sorghum flour is usually not milled centrally and fortification at the community level can be problematic. Therefore, iron fortification of sorghum flour is still in the early stages and knowledge on appropriate iron compounds for sorghum fortification is scarce. Cereals such as sorghum are particularly difficult to fortify with iron, since they contain significant amounts of both PA and PPs [25]. NaFeEDTA is recommended as a fortificant in cereals containing high amounts of PA [26], however, results from previous studies in tea have raised questions about its ability to overcome the inhibitory effect of PPs [27]. The usefulness of NaFeEDTA in the presence of PPs from other sources is unknown. Other approaches to enhance iron bioavailability from sorghum flour would be the addition of ascorbic acid (AA), a well-known iron absorption enhancer [28], or the enzymatic reduction of inhibitory PPs by a polyphenol oxidase (PPO). Phytase has been used in a similar way to degrade PA and increase iron absorption from cereal meals [25].

Here we report three stable isotope absorption studies in young women consuming dephytinized iron-fortified porridges made from white and brown sorghum flour containing different concentrations of PPs. The studies investigated a potential dose-dependent inhibitory effect of sorghum

PPs on iron absorption and the ability of NaFeEDTA, AA and laccase to overcome the inhibitory effect.

Methods

Participants

Fifty, apparently healthy, non-pregnant, non-lactating women aged between 18 and 40 y, with a normal body mass index (18·5–25 kg/m^2) and a body weight <65 kg were recruited from among the student and staff population of ETH Zurich, Switzerland. Participants were randomly allocated to three iron absorption studies, with 16 or 18 participants per study. Women with known metabolic, chronic and gastro-intestinal disease, as well as woman on long-term medication (except oral contraceptives), were excluded from participating. Intake of vitamin and/or mineral supplements was not allowed during and two weeks before the studies. No women were recruited who had donated blood or experienced substantial blood loss within 4 months of the beginning of the study. The study was conducted according to the guidelines laid down in the Declaration of Helsinki and all procedures involving human subjects were approved by the ethical committee at ETH Zurich, Switzerland. All participants provided informed written consent.

Study design

Three iron absorption studies were conducted using a randomized cross-over design with each participant serving as her own control. In study 1, the influence of different amounts of sorghum PPs on iron absorption was investigated using white and brown sorghum flour and a mixture of both. Study 2 investigated the potential of NaFeEDTA to overcome the inhibitory effect of sorghum PPs on iron absorption and study 3 investigated the potential of ascorbic acid (AA) addition and PPO pre-treatment of sorghum meals in overcoming the inhibitory effects. In each study, three differently labeled sorghum test meals were administered to the participants on three consecutive days in a randomized fashion (**Figure 1**). Iron absorption was determined by stable isotope technique measuring the incorporation of isotopic iron labels into erythrocytes 14 d after the administration of the third test meal [29].

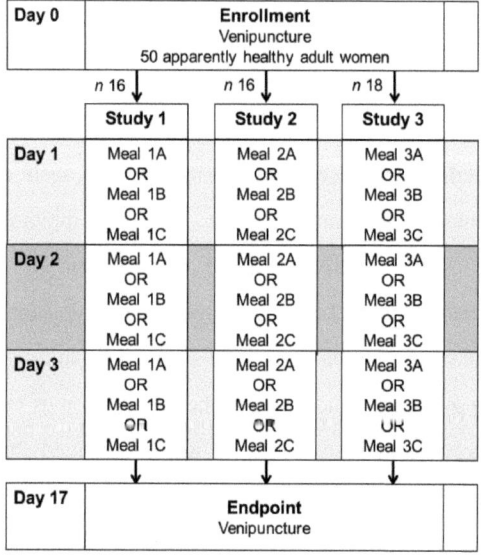

Figure 1: Schematic diagram of the study protocol. Meals 1A and 2A are low PP sorghum porridges fortified with FeSO4; meal 1B is medium PP sorghum porridge fortified with FeSO4; meals 1C, 2B and 3A are high PP sorghum porridges fortified with FeSO4; meal 2C is high PP sorghum porridge fortified with NaFeEDTA; meal 3B is high PP sorghum porridge fortified with AA and FeSO4; meal 3C is reduced PP sorghum porridge pre-treated with laccase and fortified with FeSO4. AA, ascorbic acid; PP, polyphenol.

On day 0, body weight and height were measured and the first blood sample was taken for iron status (hemoglobin, Hb; plasma ferritin, PF) and inflammation (C-reactive protein, CRP) measurements. On day 1, the first labeled test meal was administered in two servings with a 3h minimum interval. On day 2 and 3, the second and third test meals were administered after an overnight fast according to the same procedure. Test meals were fed under strictly standardized conditions and under close supervision of the investigators. Complete intake of isotopically labeled meals was assured by rinsing the test meal bowl twice with 10 ml high-purity water (18 mol/L) after consumption and letting the participants consume the rinsing water. No intake of food and fluids was allowed between the two meals and for 3h after the second meal. In all three studies, the servings of the three different test meals were labeled with 2 mg ^{54}Fe, ^{57}Fe or ^{58}Fe. Fourteen days after the third test meal (day 17), a second blood sample was drawn for iron isotopic analysis.

Test meals

A white (CSM 485; *Kossa*) and a brown sorghum (ICSV 1001; *Framida*) variety from Mali and Burkina Faso, respectively, were used to prepare test meals with low, medium and high PP concentrations. White sorghum was decorticated in Mali using an abrasive dehulling device (TADD, Venable Machine Works, Saskatoon, Canada). Brown sorghum was decorticated in France using an abrasive laboratory scale decorticator (DMS 500, Electra, Poudenas, France). Both varieties were washed before decortication and an extraction rate of 86–90% was obtained. After decortication they were milled in Switzerland using a centrifugal mill (Retsch, Haan, Germany) equipped with a titanium sieve to avoid iron contamination. Flours were prepared in bulk and used for the three studies.

In study 1, test meals were based on the white sorghum flour labeled with 2 mg ^{54}Fe as FeSO$_4$ (1A), brown sorghum flour labeled with 2 mg ^{58}Fe as FeSO$_4$ (1C) and a 1:1 mixture of the two flours labeled with 2 mg ^{57}Fe as

FeSO$_4$ (1B), respectively. In the second study, one test meal was based on the white sorghum flour labeled with 2 mg ^{54}Fe as FeSO$_4$ (2A) and the other two on the brown sorghum flour labeled with 2 mg ^{57}Fe as FeSO$_4$ (2B) or with 2 mg ^{58}Fe as NaFeEDTA (2C). In study 3, test meals were based on the brown sorghum flour labeled with 2 mg ^{58}Fe FeSO$_4$ (3A), labeled with 2 mg ^{57}Fe as FeSO$_4$ and containing 40 mg AA per serving (3B) or pre-treated with laccase and labeled with 2 mg ^{54}Fe as FeSO$_4$ (3C).

In all three studies, test meal servings (235 g) were based on 50 g sorghum flour, 15 g sugar and 170 g water. Except for the test meal pre-treated with laccase, test meals were prepared by adding sorghum flour and sugar to previously heated water (~55°C), and the pH of the resulting slurry was adjusted to 5·0–5·2 by using 0·5 M hydrochloric acid (Sigma-Aldrich, Steinheim, Germany). Approximately 3000 phytase units (FTUs)/100 g flour was added to the slurry using DSM phytase 5000 liquid provided by DSM Nutritional Products, Switzerland and the slurry was held at ~55°C for 1h in a water bath to allow for complete PA degradation. The slurry was then heated to 80°C to inactivate phytase. After cooling, water was added to adjust to the ratio of water to sorghum flour of 3·4:1 and portions corresponding to 50 g dry weight of sorghum flour were weighed as test meal servings.

The test meals pre-treated with laccase were prepared by adding sorghum flour and sugar to previously heated water (~40 °C) without adjustment of pH. Seven–10 ml phytase and 120 ml laccase (NS 33136; Novozymes, Dittingen, Switzerland) was added to the slurry and the slurry was held at ~40°C for 4h under continuous stirring in a water bath to allow for complete PA and partial degradation of PPs. The slurry was then heated to 80°C to inactivate enzymes and processed in the same way as the other test meals mentioned above.

The test meals were prepared in batches and servings were stored frozen until the day of the administration. Two mg of iron label in solution was added to each test meal serving shortly before test meal administration. Two-hundred-fifty g deionized water was consumed with each test meal serving.

Test meal analysis

Iron in the sorghum flours and test meals was analyzed by graphite-furnace atomic absorption spectrophotometry (GF-AAS, AA240Z; Varian, Palo Alto, CA) after mineralization by microwave digestion (MLS ETHOSplus, MLS GmbH; Leutkirch, Germany). For the estimation of soil contamination, aluminum and titanium in the sorghum flours were analyzed by x-ray fluorescence analysis using Spectro X-Lab 2000 (Spectro Analytical Instruments GmbH, Kleve, Germany). A matrix correction for food samples was applied. The PA concentration was measured by using a modification of the Makower method [30], in which iron was replaced by cerium in the precipitation step. After the mineralization of the precipitates, inorganic phosphate was determined according to Van Veldhoven and Mannaerts [31] and converted into PA concentrations. The total PP concentration in the sorghum flours and the test meals was determined by using a modified Folin-Ciocalteau method [32], and was expressed as gallic acid equivalents (GAEs). Wheat bran (PA assay) and milled beans (PP assay), flushed with argon to avoid PP oxidation were analyzed together with each series of samples and were used as an internal control material to monitor reproducibility. The *in vitro* accessible iron was determined using the method described in Luten et al. [33]. This method includes a gastric stage, where the sample is incubated in an HCl-pepsin solution at pH 2 and 37°C for 2h, followed by an intestinal stage determining dialyzability.

Preparation of isotopically labeled iron

Isotopically-labeled $^{54}FeSO_4$, $^{57}FeSO_4$, and $^{58}FeSO_4$ were prepared from isotopically enriched elemental iron (^{54}Fe-metal: 99·9 % enriched; ^{57}Fe-metal: 97·1 % enriched; ^{58}Fe-metal: 99·9 % enriched; all Chemgas, Boulogne, France) by dissolution in 0·1 M sulfuric acid. The solutions were flushed with argon to keep the iron in the +II oxidation state. Prepared iron tracer solutions were analyzed for iron isotopic composition and tracer iron concentration by reversed isotope dilution mass spectrometry. Na^{58}FeEDTA was prepared in solution from ^{58}Fe-enriched elemental iron (^{58}Fe-metal: 99·9 % enriched; Chemgas, Boulogne, France). The metal was dissolved in 2 mL HCl and diluted with water. The resulting $FeCl_3$ solution was mixed with a freshly prepared aqueous Na_2EDTA solution ($Na_2EDTA \cdot H_2O_2$; Sigma Chemical Co, St Louis, MO) at a molar ratio of 1:1 (Fe:EDTA). The resulting NaFeEDTA solution was added to the test meal.

Blood Analysis

Hb was measured with a Coulter Counter (AcT8 Counter; Beckman Coulter, Krefeld, Germany); anemia was defined as hemoglobin <120 g/L [34]. PF and CRP were measured with an IMMULITE automatic system (Siemens, Zurich, Switzerland). ID was defined as PF <15 µg/L and ID anemia as Hb <120 g/L and PF <15 µg/L [34]. High-sensitivity CRP concentrations in healthy individuals were <5 mg/L.

Each isotopically enriched blood sample was analyzed in duplicate for its isotopic composition. Whole blood was mineralized by microwave digestion, and iron was separated by anion-exchange chromatography and a subsequent precipitation step with ammonium hydroxide [35]. Iron isotope ratios were determined by an MC-ICP-MS instrument (NEPTUNE, Thermo Finnigan, Bremen, Germany).

Calculation of percentage iron absorption

The percentage of ^{54}Fe, ^{57}Fe and ^{58}Fe labels in the blood were calculated based on the dose of administered isotopes (4 mg/type of test meal), the shift in iron isotope ratios and the estimated amount of iron circulating in the body. Circulating iron was calculated based on the blood volume estimated from height and weight and measured Hb concentration [36]. The calculations were based on the principles of isotope dilution, taking into account that iron isotopic labels were not monoisotopic using the method by Turnlund et al. [37]. Calculation of iron absorption is reported in detail in **Supplemental Figure 1**. For calculation of percentage absorption, 80% incorporation of the absorbed iron into red blood cells was assumed [38].

Statistical analysis

Analyses were conducted with SPSS statistical software (SPSS 19·0; SPSS Inc.). All values were converted to their logarithms for statistical analysis and reconverted for reporting. Test meal analyses were reported as means ± SD. Results of iron absorption are presented as geometric means with the 95% CI in parentheses. Results of iron status and inflammation measures were presented as means ± SD if normally distributed. Medians with 25th–75th percentile ranges in parentheses were used to present results which were not normally distributed after log conversion (CRP, PF). Iron absorption from the different test meals within the same participant was compared by a linear mixed model with repeated measure followed by a post hoc Bonferroni test for multiple comparisons. In the linear mixed model, types of test meal and study day were the fixed effects and individual participant was the random effect. The dependent variable was iron absorption and study day was treated as the repeated measure assuming an unstructured covariance type. For the between-study comparison, the same linear mixed model with repeated measure was applied but study as a fixed effect was added and plasma ferritin as a covariate. A 1-way ANOVA followed by a post hoc Bonferroni test for multiple comparisons was used for participant's baseline characteristics as well as for test meal composition (total iron, PPs). Differences were considered significant at $P<0\cdot05$. The studies were powered to detect an intra-subject difference of 30% in iron absorption with α level of $0\cdot05$.

Results

Participant characteristics

Six participants (two in study 1, three in study 2, one in study 3) had ID and 3 were anemic (one in each study). Two participants had ID anemia (one in study 1 and 2). Three women had a slightly elevated CRP concentration of >3 mg/L, but none had a CRP concentration >5 mg/L. None of the variables in **Table 1**, except age, differed significantly between study groups.

Table 1: Age, anthropometric features and Hb, PF, and CRP concentrations of participating healthy adult women at baseline
(Mean values, standard deviations, medians and 25th–75th percentiles)

	Study 1 (n 16)		Study 2 (n 16)		Study 3 (n 18)	
	Mean	SD	Mean	SD	Mean	SD
Age (y)	24·9	2·8[a]	25·7	3·2[a]	22·1	3·0[b]
Weight (kg)	59·2	5·4	59·3	3·0	57·8	5·7
Height (cm)	164	7	166	5	166	6
BMI (kg/m^2)	22·0	2·1	21·7	1·9	21·4	1·9
Hb (g/L)	137	10	135	8	135	7
PF (μg/L)						
Median	57·8		31·4		43·2	
25th, 75th percentiles	36·6, 86·6		20·6, 45·1		27·8, 52·0	
Plasma CRP (mg/L)						
Median	0·4		1·3		0·3	
25th, 75th percentiles	0·2, 0·9		0·2, 2·0		0·2, 0·6	

CRP, C-reactive protein; Hb, hemoglobin; PF, plasma ferritin.
[a,b] Values within a row with unlike superscript letters were significantly different ($P<0·05$).

Test meal composition

Uncooked flours from decorticated white and brown sorghum had PP concentrations expressed as GAEs of 21 (SD 1) mg and 553 (SD 16) mg/100 g, respectively. The PP concentration of the white sorghum porridge (low PP) was ~10 and ~4 times lower than that of the brown sorghum porridge (high PP) or that of the porridge based on a mixture of white and brown sorghum (medium PP), respectively (P<0·0001) (**Table 2**). Incubation with laccase reduced PP concentration in the brown sorghum test meals by about 75% (P<0·0001). However, laccase did not completely degrade PPs and we were not able to attain a further PP reduction, as measured by Folin-Ciocalteu method by modifying incubation conditions (time, temperature, pH). PA was below the detection limit in all test meals (<8 mg/test meal serving).

The iron concentration of the white sorghum flour was 2·6 (SD 0·3) mg/100 g. The total iron concentration of the brown sorghum flour used to prepare the test meals was 5·8 (SD 0·4) mg/100 g, whereas 2·1 (SD 0·1) mg/100 g was native iron and the rest (~3·7 mg/100 g) was contamination iron. The native iron concentration of the brown sorghum was obtained by analyzing the iron concentration of brown sorghum grains which, in contrast to the grains used to obtain the bulk of flour, were carefully washed after decortication. Iron contamination of brown sorghum flour was confirmed by the small silver spots detected on the magnet stirrer after the incubation of the flour with laccase. The amount of contamination iron was ~0·9 mg/portion in the test meals based on white and brown sorghum (meal 1B) and ~1·9 mg/portion in the test meals based on brown sorghum (meals 1C, 2B, 3A). The total iron concentrations and the proportion of different iron sources in the different test meals are shown in Table 2.

The aluminum concentrations of the non-contaminated and contaminated grains were 4·9 (SD 1·4) mg/100 g and 3·0 (SD 1·6 mg)/100 g, respectively; and titanium concentrations of the non-contaminated and contaminated grains were below detection limit (0·5 mg/100 g). The *in vitro* accessible iron after *in vitro* gastric digestion followed by a dialyzability procedure was 578 (SD 87) µg/100 g and 608 (SD 30) µg/100 g from the non-contaminated and contaminated flour, respectively, indicating that the contamination iron was not soluble during simulated gastric digestion and/or dialyzable.

Table 2: Composition of different sorghum test meals based on 50 g flour and consumed by healthy adult women
(Mean values and standard deviations)

Meals	Description (iron compound)	PP* (mg/serving) Mean	PP* (mg/serving) SD	PA† (mg/serving) Mean	PA† (mg/serving) SD	Total iron‡ (mg/serving) Mean	Total iron‡ (mg/serving) SD	Native iron§ (mg/serving)	Contamination iron§ (mg/serving)	Fortification iron| (mg/serving)
1A,2A	Low PP sorghum porridge (FeSO$_4$)	17[a]	1	n.d.		3·6[a]	0·1	1·6	—	2·0
1B	Medium PP sorghum porridge (FeSO$_4$)	73[b]	1	n.d.		4·0[a,b]	0·2	1·1	0·9	2·0
1C,2B,3A	High PP sorghum porridge (FeSO$_4$)	167[c]	8	n.d.		5·0[b]	0·7	1·1	1·9	2·0
2C	High PP sorghum porridge (NaFeEDTA)	167[c]	8	n.d.		5·0[b]	0·7	1·1	1·9	2·0
3B	High PP sorghum porridge + 40 mg AA (FeSO$_4$)	167[c]	8	n.d.		5·0[b]	0·7	1·1	1·9	2·0
3C	Reduced PP sorghum porridge, pre-treated with laccase (FeSO$_4$)	42[d]	1	n.d.		5·1[b]	0·1	1·1	1·9	2·0

AA, ascorbic acid; PA, phytic acid; PP, polyphenol.
[a,b,c,d] Mean Values within a column with unlike superscript letters were significantly different (PP=$P<0.0001$; iron=$P<0.05$)
* $n = 3$ independent analyses. Expressed as gallic acid equivalents.
† n.d., below the limit of detection (<8 mg/serving), $n = 3$ independent analyses.
‡ $n = 3$ independent analyses in prepared sorghum porridges.
§ Calculation of native and contamination iron is based on measured iron concentrations in flour of washed (non-contaminated) and unwashed (contaminated) brown sorghum grains; 2·1 ± 0·1 mg/100 g and 5·8 ± 0·4 mg/100 g, respectively.
\| 2 mg iron added as ^{54}Fe, ^{57}Fe or ^{58}Fe in form of FeSO$_4$ or NaFeEDTA.

Table 3: Percentage iron absorption by healthy adult women from different sorghum test meals fortified with ferrous sulfate or sodium iron EDTA
(Mean values and 95% confidence interval)

Study no.	n	Meals	Iron absorption (%)	
			Geometric mean	95% CI
1	16	1A) Low PP sorghum porridge (FeSO$_4$)	8·5^a	5·4, 13·4
		1B) Medium PP sorghum porridge (FeSO$_4$)	3·2^b	1·7, 5·9
		1C) High PP sorghum porridge (FeSO$_4$)	2·7^b	1·5, 5·0
2	16	2A) Low PP sorghum porridge (FeSO$_4$)	10·7^a	7·2, 15·9
		2B) High PP sorghum porridge (FeSO$_4$)	2·7^b	1·5, 5·0
		2C) High PP sorghum porridge (NaFeEDTA)	4·6^c	2·6, 8·0
3	18	3A) High PP sorghum porridge (FeSO$_4$)	4·6^a	3·3, 6·5
		3B) High PP sorghum porridge + AA (FeSO$_4$)	13·6^b	9·9, 18·7
		3C) Reduced PP sorghum porridge, pre-treated with laccase (FeSO$_4$)	4·8^a	3·5, 6·6

AA, ascorbic acid; PP, polyphenol.
^{a,b,c} Geometric mean values within studies 1, 2 and 3 with unlike superscript letters were significantly different at $P<0·001$ (except meal 2B vs. 2C, $P<0·05$).

Iron absorption measurements

In study 1, mean percentage iron absorption from the medium and high PP sorghum meals was ~3 times lower ($P<0·001$) than that from low PP sorghum meals **(Table 3)**. There was no difference in iron absorption ($P=0·9$) between the medium PP and the high PP sorghum meal. In study 2, iron absorption from FeSO$_4$-fortified low PP sorghum meals was 4 and 2.3 times higher ($P<0·001$) than that from FeSO$_4$- or NaFeEDTA-fortified high PP sorghum meals, respectively. However, iron absorption from high PP sorghum meals fortified with NaFeEDTA was almost 2-fold higher than from high PP sorghum meals fortified with FeSO$_4$ ($P<0·001$). In study 3, mean percentage iron absorption from the high PP sorghum meals with added AA was almost 3-fold higher than that from the high PP sorghum meals without AA ($P<0·001$). Decreasing PPs of high PP sorghum meals by laccase pre-treatment did not increase iron absorption compared with high PP sorghum meals without laccase pre-treatment ($P=0.4$). Regarding between-study comparison (**Figure 2**), iron absorption from meals containing 167 mg PPs per serving fortified either with $FeSO_4$ (1C, 2B, 3A) or NaFeEDTA (2C) were different ($P<0.05$), but absorption from the latter was not different from the $FeSO_4$-fortified meals containing 72 mg PPs ($P=0.8$) or the laccase pre-treated meals (42 mg PPs; $P=0.06$). Between-study comparison of meals based on white sorghum (meals 1A, 2A) and the AA-fortified meal based on brown sorghum (3B) showed no difference in iron absorption ($P=1.0$).

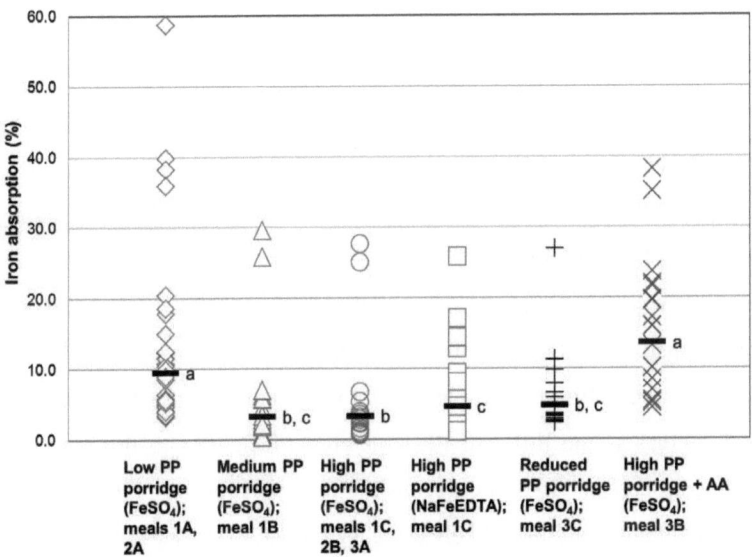

Figure 2: Between-study comparison of iron absorption from the different types of sorghum porridges consumed by healthy adult women. Values are individual data points with the horizontal bar representing the geometric mean. Absorption data from the same type of meals administered in the different studies (meals 1A and 2A, n 32; meals 1C, 2B and 3A, n 50) were combined for statistical analysis. Porridges with unlike letter were significantly different P<0•05. AA, ascorbic acid; PP, polyphenol.

Discussion

In study 1, we have shown that 73 mg and 167 mg PPs from brown sorghum, expressed as GAEs, significantly decrease absorption of fortification iron from standardized dephytinized sorghum porridges. These results indicate that the iron bioavailability from traditional Western African porridges based on red or brown sorghum, containing 80–160 mg PPs [9], would be expected to be impaired even if most of the PA is degraded by fermentation. However, it should be noted that we used standardized dephytinized test meals, and the effect of PPs on iron absorption might be differently pronounced in composite, non-dephytinized, sorghum-based meals. In a study with beans, the inhibitory effect of PPs was no longer significant from composite bean meals containing different enhancers and inhibitors [39].

Brown sorghum PPs would seem to be slightly less inhibitory than the PPs of black tea, since Hurrell et al reported that 100 and 200 mg GAEs of black tea PPs reduced iron absorption from an iron-fortified bread meal by about 80% [19]. The inhibitory effect of sorghum PPs in the present study was similar to the inhibitory effect of PPs from herb tea [19] but markedly greater than the inhibitory effect of PPs from common beans [40]. The reason for these differences in the inhibitory potential of PPs from different food sources is the different PP structures which lead to different iron binding properties [20, 21]. The responsible structures for the inhibitory effect seen in our study are most likely catechol and galloyl groups of condensed tannins which represent 72% of the PP compounds found in the brown sorghum variety used in our study (*Framida*) [7]. Red or brown sorghum has been reported to contain high amounts of catechols and galloyls (1350 and 550 mg/100 g grain, respectively) [17] which can form strong, nonabsorbable, complexes with iron in the intestinal tract [19, 20]. The fact that the iron absorption from the medium PP sorghum meal was as low as from the high PP sorghum meal suggests that relatively low amounts of these condensed tannins from *Framida* sorghum can strongly inhibit iron absorption. The white sorghum contained very little PPs and did not have a pigmented testa indicating that it does not contain condensed tannins. We assume that there was little or no significant interference with iron absorption from white sorghum porridges.

The results of our study are in agreement with a previous radioisotope study reporting a 2-fold higher iron absorption from a low PP sorghum than from a high PP sorghum [41], but contradictory to an early radioisotope study reporting similar iron absorption from a low and a high tannin sorghum variety [42]. In both studies, the different PA concentrations of the two sorghum varieties make interpretation difficult. More recently, Hurrell et al. reported that dephytinization of low-tannin sorghum meals increased iron absorption 2-fold but that no improvement in iron absorption was observed after dephytinization of high-tannin sorghum meals [25]. In all three previous studies, the exact concentrations of PPs in the test meals are unknown because PP measurements were made before cooking which is known to significantly decrease PP concentrations [43, 44].

The results of our second study using NaFeEDTA confirmed the earlier reports that EDTA cannot completely protect iron from PP-binding [27, 45]. Presumably at the pH of the stomach, PPs (but not PA) can still strongly compete with EDTA for iron binding. Sorghum foods are usually not completely dephytinized and contain considerable amounts of PA [44]. NaFeEDTA is known to overcome the inhibitory effect of PA [27], thus it seems a suitable compound for the fortification of sorghum flour containing both inhibitors. A previous study using NaFeEDTA as a fortificant for fermented red sorghum flour with moderate concentrations of PPs and PA reported high iron absorption of ~18% in women treated for malaria [46].

The enhancing effect of AA on iron absorption in study 3, completely overcoming the inhibitory effect of the PPs, can be attributed to its ability to reduce ferric to ferrous iron in the stomach and duodenum, and to form soluble complexes with ferric and ferrous iron at higher pH in the intestine [47]. Iron that is bound to AA can no longer bind to PPs and ferrous iron forms much weaker complexes with PPs than ferric iron [48]. Our results are in agreement with a previous study which showed that AA prevents the dose-dependent inhibitory effect of the PP compound tannic acid [49]. Adding AA to high PP sorghum foods fortified with iron would appear to be a useful approach to increase iron bioavailability, but heat and oxidation sensitivity of AA is a disadvantage and has to be considered.

It was somewhat surprising that the enzymatic reduction in total PPs in sorghum porridge by laccase did not increase iron absorption. An enhanced absorption was predicted by an *in vitro* study which reported increased *in vitro* accessible iron after PP reduction in dephytinized high tannin sorghum [50]. The *in vitro* study, however, used tyrosinase instead of laccase as the PPO, and whole red sorghum grains with a much higher initial PP concentration than the decorticated sorghum in our study. We used laccase as PPO because, in terms of PP degradation, tyrosinase did not perform well in preliminary trials. Furthermore, tyrosinase is not suitable for human food applications as there is currently no food grade product available. Food-grade laccase is an enzyme that catalyzes the oxidation of a wide range of phenolic substrates including *ortho*- and *para*-diphenols, and triphenols. Tyrosinase preferably oxidizes monophenols and a few diphenols [51]. We had assumed laccase would oxidize the catechol-groups and, as those groups are suspected to bind iron, increase absorption.

However, our results indicate that the laccase did not or only partly degrade the inhibitory PPs in the brown sorghum porridge. It seems that the remaining unchanged PPs (42 mg/serving) are still strongly inhibitory, and/or that the transformed PPs can still bind iron. Although the present study did not find laccase useful to improve iron absorption, more systematic studies with laccase and other PPOs are needed to evaluate further possibilities.

Some test meals in our study contained contamination iron (~23% and ~38% of total iron). We can exclude that the iron comes from soil contamination because grains were carefully washed before decortication, and because soil contaminated grains would have higher aluminum and titanium concentrations than that found in our grains [52]. Contamination from milling can also be excluded because the centrifugal mill was constructed with titanium. The iron contamination originates, therefore, from the decorticator made of stainless steel. Based on our results of simulated gastric digestion followed by dialysis, which showed that the quantity of *in vitro* accessible iron was identical from non-contaminated and contaminated sorghum grains, we conclude that the iron from stainless steel is not bioavailable or at best poorly bioavailable. Nevertheless, if significant amounts of the contaminant steel iron have dissolved and would have entered the common iron pool in our studies, our data for percentage iron absorption would slightly underestimate the inhibitory effect of the PPs because the additional bioavailable iron from steel would lead to a lower molar ratio of PPs to iron which could result in a higher percentage absorption from the isotopic tag. We conclude therefore that the contamination iron present in some sorghum flours would not be expected to greatly influence the measured percentage absorption values or change the conclusion on the extent to which sorghum polyphenols inhibit iron absorption.

In conclusion, our studies show that brown sorghum PPs are strong inhibitors of iron absorption. They appear to be more inhibitory than bean PPs and similar to herb teas. This means that dephyitinization of meals, based on colored sorghum varieties, is unlikely to greatly increase iron absorption as does dephyitinization of other cereals [25]. AA is the preferred enhancer of iron absorption from sorghum meals as this overcomes the inhibition of both PPs and PA, whereas the use of NaFeEDTA is much more effective against PA than PPs. Reduction of PPs by PPO appears to be possible and this treatment merits more investigation.

Acknowledgements

The authors thank Claire Monquet-Rivier (Institute de Recherche pour le Développement, Montpellier, France), Laurencia Toulsoumdé Ouattara (Centre National de la Recherche Scientifique et Technologique, Ouagadougou, Burkina Faso) and Eva Weltzien (International Crops Research Institute for the Semi-Arid Tropics, Bamako, Mali) for providing the sorghum varieties and for the decortication of sorghum. We express thanks to DSM Nutritional Products (Basel, Switzerland) and Novozymes (Dittingen, Switzerland) for donating the phytase and laccase, respectively.

The present study was supported by the INSTAPA project, which receives funding from the European Community's Seventh Framework Programme [FP7/2007–2013] under grant agreement no. 211484. The Seventh Framework Programme of the European Community had no role in the design, analysis or writing of this article.

C.I.C., I.M.E., C.Z., and R.F.H. designed research; C.I.C., I.M.E., and C.Z. conducted research; C.I.C. and C.Z. analyzed data; C.I.C., I.M.E., and

R.F.H. wrote the paper; and C.I.C., I.M.E., and R.F.H. had primary responsibility for final content. All authors read and approved the final version of the paper. None of the authors declare any conflicts of interest.

References

1. McLean E, Cogswell M, Egli I et al. (2009) Worldwide prevalence of anaemia, WHO Vitamin and Mineral Nutrition Information System, 1993-2005. *Public Health Nutr* **12**, 444-54.
2. Zimmermann MB, Hurrell RF. (2007) Nutritional iron deficiency. *Lancet* **370**, 511-20.
3. Aikawa R, Ngyen CK, Sasaki S et al. (2006) Risk factors for iron-deficiency anaemia among pregnant women living in rural Vietnam. *Public Health Nutr* **9**, 443-8.
4. Zimmermann MB, Chaouki N, Hurrell RF. (2005) Iron deficiency due to consumption of a habitual diet low in bioavailable iron: a longitudinal cohort study in Moroccan children. *Am J Clin Nutr* **81**, 115-21.
5. Rooney LW, Waniska RD (2000) Sorghum food and industrial utilization. In *Origin, History, Technology, and Production*, pp. 689-729 [Smith CW, Frederiksen RA, editors]. New York: John Wiley and Sons.
6. Godwin ID, Gray SJ (2000) Overcoming productivity and quality constraints in sorghum: the role for genetic engineering. In *Transgenic cereals*, pp. 153-77 [O'Brien L, Henry RJ, editors]. St. Paul, MN: American Association of Cereal Chemists.
7. Dicko MH, Hilhorst R, Gruppen H et al. (2002) Comparison of content in phenolic compounds, polyphenol oxidase, and peroxidase in grains of fifty sorghum varieties from Burkina Faso. *J Agr Food Chem* **50**, 3780-8.
8. Anglani C. (1998) Sorghum for human food - A review. *Plant Food Hum Nutr* **52**, 85-95.
9. Kayode APP, Adegbidi A, Hounhouigan JD et al. (2005) Quality of farmers' varieties of sorghum and derived foods as perceived by consumers in Benin. *Ecol Food Nutr* **44**, 271-94.
10. Badi S, Pedersen B, Monowar L et al. (1990) The nutritive value of new and traditional sorghum and millet foods from Sudan. *Plant Foods Hum Nutr* **40**, 5-19.
11. Onofiok NO, Nnanyelugo DO. (1998) Weaning foods in West Africa: Nutritional problems and possible solutions. *Food Nutr Bull* **19**.
12. Igbedioh SO, Ogbeni AO, Adole GM. (1996) Infant weaning practices of some Tiv women resident in Makurdi, Nigeria. *Nutr Health* **11**, 13-28.
13. Kayode AP, Nout MJ, Bakker EJ et al. (2006) Evaluation of the simultaneous effects of processing parameters on the iron and zinc solubility of infant sorghum porridge by response surface methodology. *J Agric Food Chem* **54**, 4253-9.
14. Mitchikpe ECS, Dossa RAM, Ategbo EAD et al. (2008) The supply of bioavailable iron and zinc may be affected by phytate in Beninese children. *J Food Compos Anal* **21**, 17-25.
15. Kayode AP, Linnemann AR, Hounhouigan JD et al. (2006) Genetic and environmental impact on iron, zinc, and phytate in food sorghum grown in Benin. *J Agric Food Chem* **54**, 256-62.
16. Makokha AO, Oniang'o RK, Njoroge SM et al. (2002) Effect of traditional fermentation and malting on phytic acid and mineral availability from sorghum (Sorghum bicolor) and finger millet (Eleusine coracana) grain varieties grown in Kenya. *Food Nutr Bull* **23**, 241-5.
17. Towo EE, Svanberg U, Ndossi GD. (2003) Effect of grain pre-treatment on different extractable phenolic groups in cereals and legumes commonly consumed in Tanzania. *J Sci Food Agric* **83**, 980-6.
18. Hallberg L, Rossander L, Skanberg AB. (1987) Phytates and the inhibitory effect of bran on iron absorption in man. *Am J Clin Nutr* **45**, 988-96.
19. Hurrell RF, Reddy M, Cook JD. (1999) Inhibition of non-haem iron absorption in man by polyphenolic-containing beverages. *Brit J Nutr* **81**, 289-95.
20. Brune M, Rossander L, Hallberg L. (1989) Iron absorption and phenolic compounds: importance of different phenolic structures. *Eur J Clin Nutr* **43**, 547-57.
21. Tuntipopipat S, Judprasong K, Zeder C et al. (2006) Chili, but not turmeric, inhibits iron absorption in young women from an iron-fortified composite meal. *J Nutr* **136**, 2970-4.

22. Hahn DH, Rooney LW. (1986) Effect of Genotype on Tannins and Phenols of Sorghum. *Cereal Chem* **63**, 4-8.
23. Dykes L, Rooney LW. (2006) Sorghum and millet phenols and antioxidants. *J Cereal Sci* **44**, 236-51.
24. Dykes L, Rooney LW, Waniska RD et al. (2005) Phenolic compounds and antioxidant activity of sorghum grains of varying genotypes. *J Agric Food Chem* **53**, 6813-8.
25. Hurrell RF, Reddy MB, Juillerat MA et al. (2003) Degradation of phytic acid in cereal porridges improves iron absorption by human subjects. *Am J Clin Nutr* **77**, 1213-9.
26. Hurrell R, Ranum P, de Pee S et al. (2010) Revised recommendations for iron fortification of wheat flour and an evaluation of the expected impact of current national wheat flour fortification programs. *Food Nutr Bull* **31**, S7-21.
27. Hurrell RF, Reddy MB, Burri J et al. (2000) An evaluation of EDTA compounds for iron fortification of cereal-based foods. *Brit J Nutr* **84**, 903-10.
28. Hallberg L, Brune M, Rossander L. (1989) Iron-Absorption in Man - Ascorbic-Acid and Dose-Dependent Inhibition by Phytate. *Am J Clin Nutr* **49**, 140-4.
29. Walczyk T, Davidsson L, Zavaleta N et al. (1997) Stable isotope labels as a tool to determine the iron absorption by Peruvian school children from a breakfast meal. *Fresen J Anal Chem* **359**, 445-9.
30. Makower RU. (1970) Extraction and Determination of Phytic Acid in Beans (Phaseolus-Vulgaris). *Cereal Chem* **47**, 288-96.
31. Van Veldhoven PP, Mannaerts GP. (1987) Inorganic and Organic Phosphate Measurements in the Nanomolar Range. *Anal Biochem* **161**, 45-8.
32. Singleton VL, Rossi JA. (1965) Colorimetry of total phenolics with phosphomolybdic-phosphotungstic acid reagents. *Am J Enol Viticult* **16**, 144-58.
33. Luten J, Crews H, Flynn A et al. (1996) Interlaboratory trial on the determination of the in vitro iron dialysability from food. *J Sci Food Agric* **72**, 415-24.
34. World Health Organization (2001) *Iron deficiency anemia: assessment, prevention and control.* Geneva: WHO.
35. Hotz K, Krayenbuehl PA, Walczyk T. (2012) Mobilization of storage iron is reflected in the iron isotopic composition of blood in humans. *J Biol Inorg Chem* **17**, 301-9.
36. Brown E, Hopper J, Jr., Hodges JL, Jr. et al. (1962) Red cell, plasma, and blood volume in the healthy women measured by radiochromium cell-labeling and hematocrit. *J Clin Invest* **41**, 2182-90.
37. Turnlund JR, Keyes WR, Peiffer GL. (1993) Isotope Ratios of Molybdenum Determined by Thermal Ionization Mass-Spectrometry for Stable-Isotope Studies of Molybdenum Metabolism in Humans. *Anal Chem* **65**, 1717-22.
38. Barrett JF, Whittaker PG, Williams JG et al. (1992) Absorption of non-haem iron in normal women measured by the incorporation of two stable isotopes into erythrocytes. *Clin Sci (Lond)* **83**, 213-9.
39. Petry N, Egli I, Gahutu JB et al. (2012) Stable Iron Isotope Studies in Rwandese Women Indicate That the Common Bean Has Limited Potential as a Vehicle for Iron Biofortification. *J Nutr* **142**, 492-7.
40. Petry N, Egli I, Zeder C et al. (2010) Polyphenols and phytic acid contribute to the low iron bioavailability from common beans in young women. *J Nutr* **140**, 1977-82.
41. Gillooly M, Bothwell TH, Charlton RW et al. (1984) Factors Affecting the Absorption of Iron from Cereals. *Brit J Nutr* **51**, 37-46.
42. Radhakrishnan MR, Sivaprasad J. (1980) Tannin Content of Sorghum Varieties and Their Role in Iron Bioavailability. *J Agr Food Chem* **28**, 55-7.
43. Awika JM, Dykes L, Gu L et al. (2003) Processing of sorghum (Sorghum bicolor) and sorghum products alters procyanidin oligomer and polymer distribution and content. *J Agr Food Chem* **51**, 5516-21.
44. Kayode APP, Linnemann AR, Nout MJR et al. (2007) Impact of sorghum processing on phytate, phenolic compounds and in vitro solubility of iron and zinc in thick porridges. *J Sci Food Agric* **87**, 832-8.
45. Macphail AP, Bothwell TH, Torrance JD et al. (1981) Factors Affecting the Absorption of Iron from Fe(III)EDTA. *Brit J Nutr* **45**, 215-27.

46. Cercamondi CI, Egli IM, Ahouandjinou E et al. (2010) Afebrile Plasmodium falciparum parasitemia decreases absorption of fortification iron but does not affect systemic iron utilization a double stable-isotope study in young Beninese women. *Am J Clin Nutr* **92**, 1385-92.
47. Rossander-Hulthén L, Hallberg L (1996) Dietary factors influencing iron absorption-an overview. In *Iron nutrition in health and disease*, [Hallberg L, editor]. London: John Libbey & Company.
48. Perron NR, Brumaghim JL. (2009) A Review of the Antioxidant Mechanisms of Polyphenol Compounds Related to Iron Binding. *Cell Biochem Biophys* **53**, 75-100.
49. Siegenberg D, Baynes RD, Bothwell TH et al. (1991) Ascorbic-Acid Prevents the Dose-Dependent Inhibitory Effects of Polyphenols and Phytates on Nonheme-Iron Absorption. *Am J Clin Nutr* **53**, 537-41.
50. Matuschek E, Towo E, Svanberg U. (2001) Oxidation of polyphenols in phytate-reduced high-tannin cereals: Effect on different phenolic groups and on in vitro accessible iron. *J Agr Food Chem* **49**, 5630-8.
51. Aniszewski T, Lieberei R, Gulewicz K. (2008) Research on Catecholases, Laccases and Cresolases in Plants. Recent Progress and Future Needs. *Acta Biol Cracov Bot* **50**, 7-18.
52. Stangoulis J, Sison C (2008) Crop Sampling Protocols for Micronutrient Analysis. HarvestPlus Technial Monograph Series. http://www.ifpri.org/sites/default/files/publications/tech07.pdf (accessed January 2013)

Online Supporting Material

$$R_1 = \frac{^{56}h_{nat} \cdot n_{nat} + {}^{56}h_A \cdot n_A + {}^{56}h_B \cdot n_B + {}^{56}h_C \cdot n_C}{^{54}h_{nat} \cdot n_{nat} + {}^{54}h_A \cdot n_A + {}^{54}h_B \cdot n_B + {}^{54}h_C \cdot n_C} \quad (1)$$

$$R_2 = \frac{^{56}h_{nat} \cdot n_{nat} + {}^{56}h_A \cdot n_A + {}^{56}h_B \cdot n_B + {}^{56}h_C \cdot n_C}{^{57}h_{nat} \cdot n_{nat} + {}^{57}h_A \cdot n_A + {}^{57}h_B \cdot n_B + {}^{57}h_C \cdot n_C} \quad (2)$$

$$R_3 = \frac{^{56}h_{nat} \cdot n_{nat} + {}^{56}h_A \cdot n_A + {}^{56}h_B \cdot n_B + {}^{56}h_C \cdot n_C}{^{58}h_{nat} \cdot n_{nat} + {}^{58}h_A \cdot n_A + {}^{58}h_B \cdot n_B + {}^{58}h_C \cdot n_C} \quad (3)$$

The total circulating iron is the sum of the circulating natural iron and of the circulating isotopic labels:

$$n_{nat} + n_A + n_B + n_C = N_{tot} \quad (4)$$

Equations (1), (2), (3) and (4) can be rearranged in the following form:

$(R_1 \cdot {}^{54}h_{nat} - {}^{56}h_{nat}) \cdot n_{nat} + (R_1 \cdot {}^{54}h_A - {}^{56}h_A) \cdot n_A + (R_1 \cdot {}^{54}h_B - {}^{56}h_B) \cdot n_B + (R_1 \cdot {}^{54}h_C - {}^{56}h_C) \cdot n_C = 0$

$(R_2 \cdot {}^{57}h_{nat} - {}^{56}h_{nat}) \cdot n_{nat} + (R_2 \cdot {}^{57}h_A - {}^{56}h_A) \cdot n_A + (R_2 \cdot {}^{57}h_B - {}^{56}h_B) \cdot n_B + (R_2 \cdot {}^{57}h_C - {}^{56}h_C) \cdot n_C = 0$

$(R_3 \cdot {}^{58}h_{nat} - {}^{56}h_{nat}) \cdot n_{nat} + (R_3 \cdot {}^{58}h_A - {}^{56}h_A) \cdot n_A + (R_3 \cdot {}^{58}h_B - {}^{56}h_B) \cdot n_B + (R_3 \cdot {}^{58}h_C - {}^{56}h_C) \cdot n_C = 0$

$n_{nat} + n_A + n_B + n_C = N_{tot}$

After substitution (i.e. a_{11} for $(R_1 \cdot {}^{54}h_{nat} - {}^{56}h_{nat})$), the set of equations can be written as:

$a_{11} n_{nat} + a_{12} n_A + a_{13} n_B + a_{14} n_C = b_1$
$a_{21} n_{nat} + a_{22} n_A + a_{23} n_B + a_{24} n_C = b_2$
$a_{31} n_{nat} + a_{32} n_A + a_{33} n_B + a_{34} n_C = b_3$
$a_{41} n_{nat} + a_{42} n_A + a_{43} n_B + a_{44} n_C = b_4$

Supplemental Figure 1: Calculation of iron absorption. The measured isotopic ratios R_i can be expressed in the form of equations (1), (2) and (3), where n_m are the amounts in mol of the natural iron (m=nat), the ^{54}Fe (m=A), ^{57}Fe (m=B) and ^{58}Fe (m=C) labels circulating in the blood, respectively, and $^j h_m$ is the isotopic abundance in % mol in the natural iron and the isotopic labels, with j being the mass number of the respective iron isotope.

Manuscript 3 – Complementary Food Fortificant Study

Iron bioavailability from a lipid-based complementary food fortificant mixed with millet porridge can be optimized by adding phytase and ascorbic acid but not by using a mixture of ferrous sulfate and sodium iron EDTA

Colin I Cercamondi[1], Ines M Egli[1], Evariste Mitchikpe[2], Felicien Tossou[3], Joamel Hessou[2], Christophe Zeder[1], Joseph D Hounhouigan[2] and Richard F Hurrell[1]

[1] Laboratory of Human Nutrition, Institute of Food, Nutrition and Health, ETH Zurich, Zurich, Switzerland; [2] Laboratoire de Physiologie de la Nutrition, Université d'Abomey Calavi, Benin; [3] Zone Sanitaire Natitingou, Ministère de la Santé, Benin

Journal of Nutrition 2013; 143: 1233–39

Supported by the INSTAPA project, which receives funding from the European Community's Seventh Framework Programme [FP7/2007–2013] under grant agreement no. 211484

Abstract

Home fortification with lipid-based nutrient supplements (LNSs) is a promising approach to improve bioavailable iron and energy intake of young children in developing countries. To optimize iron bioavailability from an LNS termed complementary food fortificant (CFF), 3 stable isotope studies were conducted in 52 young Beninese children. Test meals consisted of millet porridge mixed with CFF and ascorbic acid (AA). Study 1 compared iron absorption from $FeSO_4$-fortifed meals with meals fortified with a mixture of $FeSO_4$ and NaFeEDTA. Study 2 compared iron absorption from $FeSO_4$-fortifed meals without or with extra AA. Study 3 compared iron absorption from $FeSO_4$-fortified meals with meals containing phytase added prior to consumption, once without or once with extra AA. Iron absorption was measured as erythrocyte incorporation of stable isotopes. In study 1, iron absorption from $FeSO_4$ (8.4%) was higher than from the mixture of NaFeEDTA and $FeSO_4$ (5.9%; $P < 0.05$). In study 2, the extra AA increased absorption (11.6%) compared with the standard AA concentration (7.3%; $P < 0.001$). In study 3, absorption from meals containing phytase without or with extra AA (15.8 and 19.9%, respectively) increased compared with meals without phytase (8.0%; $P < 0.001$). The addition of extra AA to meals containing phytase increased absorption compared with the test meals containing phytase without extra AA ($P < 0.05$). These findings suggest that phytase and AA, and especially a combination of the two, but not a mixture of $FeSO_4$ and NaFeEDTA would be useful strategies to increase iron bioavailability from a CFF mixed with cereal porridge.

Introduction

Iron deficiency (ID) with or without anemia is common in sub-Saharan Africa, particularly in children <5 y of age with high iron requirements during growth (1). The etiology of ID in children living in developing countries is multifactorial, but major contributors are low dietary iron bioavailability and intake from complementary foods mainly based on cereals (2). Complementary foods in developing countries are often bulky, cereal-based porridges that are not only low in bioavailable micronutrients such as iron but also low in energy density. Therefore, young children in developing countries often have difficulties meeting their daily energy requirements (3).

The use of fortified complementary foods to prevent ID is recommended in guidelines for complementary feeding (4). Complementary fortified foods can be divided into commercially fortified blended foods (FBFs) and complementary food supplements such as lipid-based nutrient supplements (LNSs) or micronutrient powders (MNPs) (5). Complementary food supplements (also called *in-home* fortification of complementary foods) could be a promising approach to improve micronutrient status in younger children when parents cannot afford commercial FBFs (6). They can provide the necessary amounts of micronutrients needed by each age subgroup (e.g. 6–12 or 12–36 mo) irrespective of how much food the children eat and without major changes in food habits. LNSs have the advantage that they provide energy, essential fatty acids, and often protein. In a 6-mo intervention trial, LNSs, MNPs and iron tablets significantly increased iron status compared with a control group, but only LNSs improved the growth of young children (7, 8). Initially, LNSs also called ready-to-use foods were designed as therapeutic foods to treat severe malnutrition (9–11). The use to prevent malnutrition and promote growth came into focus over recent years (12).

For the present study, a LNS type named complementary food fortificant (CFF) was developed. The CFF is in the form of a paste and contains a mixture of micro- and macronutrients which should be added as a fortificant to cereal-based porridges commonly consumed by children in developing countries. The CFF was designed as a preventive food with the capability to treat mild to moderate malnutrition. The recommended portion of CFF (30 g) will thus provide lower amounts of fat than previous therapeutic LNSs and the levels of iron and other micronutrients are adjusted so as to prevent deficiencies and treat mild to moderate malnutrition. Regarding iron fortification level, 10% iron bioavailability from the cereal-based porridge mixed with the CFF was assumed; so for the CFF containing 6 mg iron, children 12–36 mo old would be able to meet their daily iron requirements of 0.6 mg (1) by consuming one recommended CFF portion per day.

To test and optimize iron absorption from an iron and ascorbic acid (AA) fortified CFF blended with millet porridge, we carried out 3 iron absorption studies in young Beninese children using a stable isotope technique. In the first study, we compared iron absorption from $FeSO_4$ to a 1:1 mixture of NaFeEDTA and $FeSO_4$. In the second study, we investigated the effect of extra AA on iron absorption from $FeSO_4$. In study 3, we investigated the effect of adding phytase just before consumption on iron absorption; once with extra AA, once without.

Subjects and Methods

Subjects

Sixty apparently healthy children, 19–36 mo of age, of both sexes were recruited in Natitingou, Atacora Department, Benin. Children and their caregivers were registered during meetings in different districts of Natitingou and invited for the screening of exclusion criteria at the local state hospital. Exclusion criteria included fever (ear temperature >37.5°C), symptomatic malaria (asexual *P. falciparum* parasitemia in blood smears + fever), body weight <8.3 kg, weight for age z-score < –3, sickle cell trait [hemoglobin (Hb) S], severe hematuria, regular intake of medications and known gastrointestinal or metabolic disorders. No children were recruited who had received a blood transfusion or experienced substantial blood loss within 6 mo of the beginning of the study. None of the children consumed vitamin or mineral supplements and fortified foods 2 wk before and during the entire study. Ethical approval for the study was given by the ethical review committee at the Ministry of Health in Benin and ETH Zurich (Zurich, Switzerland). The caregivers of the children were informed about the aims and procedures of the study, and oral and written consent was obtained.

Study design

The recruited children were randomly allocated to the 3 studies. In all studies, a randomized cross-over design was used and each child served as his own control. Two different test meals (A and B) on 2 consecutive days were fed to each child in study 1 and 2. Three different test meals (A, B and C) on 3 consecutive days were fed to each child in study 3. The order of the different test meals was randomized in all studies. Test meals were fed in the morning after a fast of at least 2 h. No intake of food and fluids was allowed during the 3 h following the intake of the labeled test meals. During recruitment (baseline measurements), body weight and height of the children were measured and a first blood sample was drawn for iron status determination [Hb, plasma ferritin (PF), C-reactive protein (CRP)], detection of Hb S, and malaria parasitemia. Stool and urine samples were taken for the detection of soil transmitted helminthes. Fourteen days after the last test meal (endpoint measurements), body weight and height were measured again and a second blood sample was drawn for iron status determination and iron isotopic analysis. Stool and urine samples were taken again for confirmation of soil transmitted helminthes analyses. Ear temperature was measured at screening and endpoint, and daily before administration of test meals. Iron absorption was determined by using a stable isotope technique in which the incorporation into erythrocytes of isotopic iron labels was measured 14 d after the administration of the last test meal (13).

Tested enhancers of iron absorption

The principle of a CFF is to add a micronutrient mixture to the CFF so as to meet the requirements of young children for multiple micronutrients. In the current test phase of our CFF such a micronutrient mixture was not yet used. However, AA would almost certainly be part of the micronutrient mixture so as to cover the child's requirements and also enhance iron absorption. In all 3 studies, 40 mg AA [L(+)-ascorbic acid puriss; Sigma-Aldrich] diluted in 1 ml distilled water was added to all test meals shortly before consumption. In study 2 and 3, meals 2B and 3C contained 40 mg extra AA (total 80 mg AA).

In study 1, test meal 1B was fortified with NaFeEDTA according to the acceptable daily intake (ADI) of 1.9 mg EDTA/kg body weight per day (14). To meet ADI regulations, a 1:1 mixture of $FeSO_4$ and NaFeEDTA was used and children with a body weight below 8.3 kg who would consume EDTA in excess were excluded.

In study 3, DSM phytase Tolerase^{TM} 20000G provided by DSM Nutritional Products, Switzerland was added to test meals 3B and 3C at room temperature just before consumption. The quantity of phytase added to the meals was calculated assuming the following: a phytate concentration of 200 mg/portion and a residence time in the stomach of 60 min. Phytase activity is measured as the amount of enzyme that liberates 1 μmol inorganic phosphorus/min and is called a phytase unit (FTU). Because the phytase activity at gastric pH is expected to be 50–60%, 8 FTU was necessary to adequately degrade 1 μmol phytate (~0.7 mg phytate), leading to ~40 FTU per serving to completely degrade phytate in the stomach. To ensure complete degradation, we used a factor of 10 and added 400 FTU phytase which is in the same range as in a previous efficacy study (15).

Stable isotope labels

Isotopically-labeled $^{54}FeSO_4$, $^{57}FeSO_4$, and $^{58}FeSO_4$ were prepared from isotopically enriched elemental iron (^{54}Fe-metal: enriched >99.8%; ^{57}Fe-metal: enriched >97.8%; ^{58}Fe-metal: enriched >99.8%; all Chemgas) by dissolution in 0.1 M sulfuric acid. The solutions were flushed with argon to keep the iron in the +II oxidation state. Na^{58}FeEDTA was prepared in solution from ^{58}Fe-enriched elemental iron (^{58}Fe-metal: 99.9 % enriched; Chemgas). The metal was dissolved in 2 mL HCl and diluted with water. The resulting $FeCl_3$ solution was mixed immediately before use with an aqueous Na$_2$EDTA solution (Na$_2$EDTA · H$_2$O$_2$; Sigma Chemical) at a molar ratio of 1:1 (Fe:EDTA). The resulting NaFeEDTA solution was added to the test meal.

All test meals were fortified with 6 mg iron and labeled as follows: In study 1, meal 1A, 3 mg ^{58}Fe as FeSO$_4$ plus 3 mg iron as ^{56}FeSO$_4$ (normal isotopic composition); and meal 1B, 3 mg ^{57}Fe as FeSO$_4$ plus 3 mg ^{57}Fe as NaFeEDTA. Study 2 was: meal 2A, 3 mg ^{58}Fe as FeSO$_4$ plus 3 mg iron as ^{56}FeSO$_4$; and meal 2B, 3 mg ^{54}Fe as FeSO$_4$ plus 3 mg iron as ^{56}FeSO$_4$. In study 3, labels were: meal 3A, 3 mg ^{58}Fe as FeSO$_4$ plus 3 mg iron ^{56}FeSO$_4$; meal 3B, 3 mg ^{57}Fe as FeSO$_4$ plus 3 mg iron ^{56}FeSO$_4$; and meal 3C, 3 mg ^{54}Fe as FeSO$_4$ plus 3 mg iron as ^{56}FeSO$_4$.

Test meals

The test meal consisted of 80 g of millet porridge blended with 30 g CFF. Pearl millet (*Pennisetum glaucum*) bought from the local market in Natitingou was manually cleaned and washed and then sun-dried. Dried millet was milled using a portable household mill (Novum, Hawos Kornmühlen). To obtain a fine flour, a sieving step was added after milling. The flour was produced in bulk and used for the entire study. The millet porridge was prepared freshly on each study day by weighing and mixing millet flour and water at a ratio of 1:2.2. This blend was then added to boiling water (blend: boiling water ~1:4.1) and cooked for 1 min. After cooling, the porridge was blended with the CFF. The CFF was based on 24% canola oil (Florin), 23% peanut paste (Central African Seed Services), 23% soybean flour from milled extruded soybean flakes (Morga), 22% icing sugar (Zuckermühlen Rupperswil) and 8% palm stearin (Florin). Ingredients were blended at ETH Zurich, Switzerland using a food mixer

(Kenwood Swiss). The CFF was produced in bulk and used for the entire study.

Weighed amounts of labeled and unlabeled iron as $FeSO_4$ and NaFeEDTA in dilute acid and AA and phytase were blended with the millet porridge and the CFF shortly before feeding. The children consumed the test meals with the assistance of their caregivers under close supervision of investigators in a hospital facility in Natitingou. Complete intake of isotopically labeled meals was assured by rinsing the bowl 2 times with 10 ml distilled water after consumption of the test meal and letting the children consume the rinsing.

Test meal analysis

Sample analyses were done in triplicate at ETH Zurich, Switzerland. Iron concentrations of millet porridge and CFF were measured by graphite-furnace atomic absorption spectrophotometry (AA240Z; Varian) after mineralization by microwave digestion (MLS ETHOSplus, MLS). The phytate concentration was measured by using a modification of the Makower method (16), in which iron was replaced by cerium in the precipitation step. After the mineralization of the precipitate, inorganic phosphate was determined according to Van Veldhoven and Mannaerts (17) and converted into phytate concentrations. Total polyphenol concentration in the test meal was determined by using a modified Folin-Ciocalteau method, as suggested by Singleton et al. (18), and was expressed as gallic acid equivalents (GAEs).

Blood Analysis

Blood smears were screened for Hb S by using the bisulfite method (19) at the hospital in Natitingou, Benin. Hb was measured in whole blood on the day of collection by using HemoCue hemoglobin 201+ (HemoCue); anemia was defined as Hb <11 g/dL (1). PF and CRP were measured with an IMMULITE automatic system (Siemens) at ETH Zurich, Switzerland. ID was defined as PF <12 µg/L and ID anemia as Hb <11 g/dL and PF <12 µg/L (1). Expected high-sensitivity CRP concentrations for healthy individuals were <5 mg/L.

Each isotopically enriched blood sample was analyzed in duplicate for its isotopic composition. Whole blood was mineralized by microwave digestion and iron was separated by anion-exchange chromatography and a subsequent precipitation step with ammonium hydroxide (20). Iron isotope ratios were determined by an MC-ICP-MS instrument (NEPTUNE, Thermo Finnigan) at ETH Zurich, Switzerland.

Parasite diagnosis

Thick and thin blood smears were stained in duplicate by using the Giemsa coloration technique and were examined independently by 2 experienced microscopists (21). Rapid malaria diagnostic tests (ICT Mlo1 Malaria Pf Kit; ICT Diagnostics) were used to support the results of the blood smears.

For the detection of soil transmitted helminthes the KatoKatz method and the syringe filtration technique were used to examine stool and urine samples (21). Hematuria was detected using test strips for urinalysis (Hemastix, Siemens).

Calculation of iron absorption

The amounts of ^{54}Fe, ^{57}Fe and ^{58}Fe labels in the blood were calculated based on the shift in iron isotope ratios and the estimated amount of iron circulating in the body. Circulating iron was calculated based on the blood volume estimated from height and weight and measured Hb concentration (means from recruitment and endpoint measurements) (22). The calculations were based on the principles of isotope dilution, taking into account that iron isotopic labels were not monoisotopic, using the methods described by Walczyk et al. (13) in study 1 and 2 and by Turnlund et al. (23) in study 3. Calculation of iron absorption in study 3 is shown in detail in **Supplemental Figure 1**. For calculation of fractional absorption, 90% incorporation of the absorbed iron into RBCs was assumed (24).

Statistical analysis

Data were analyzed in Excel (Microsoft Office 2007; Microsoft) and SPSS (version 19.0; SPSS Inc.). Z-scores were calculated using WHO standards (WHO Anthro 3.2.2). Results of iron absorption, food analysis, age, anthropometric features, Hb, PF, and CRP were presented as means ± SDs if normally distributed. If not normally distributed, data were log transformed for analysis, reconverted for reporting, except for between-study comparison of iron absorption (**Fig. 1**), and presented as geometric mean values (95% CI). Iron absorption from the different test meals within the same child was compared by a linear mixed model with repeated-measure followed by a post hoc Bonferroni test for multiple comparisons. In the linear mixed model, types of test meal and study day were the fixed effects and individual child was the random effect. The dependent variable was iron absorption and study day was treated as the repeated measure assuming an unstructured covariance type. For the between-study comparison, the same linear mixed model with repeated-measure was applied, but study as a fixed effect was added and PF as covariate. Between-study comparisons of children (age, anthropometric features, Hb, PF, CRP) were done by 1-way ANOVA followed by a Bonferroni test. Differences were considered significant at $P < 0.05$. A sample size of 16 children/group is sufficient to detect an intra-subject difference of 30% in iron absorption with an α level of 0.05. To compensate for possible dropouts, ~20 children were recruited for each study.

Results

Participant characteristics

Fifty-two children (34 boys and 18 girls) completed the study, 18 in study 1 and 3, and 16 in study 2. Four children were excluded because they did not completely consume the test meals and 3 children because they either developed fever, malaria, or a wound infection requiring medication during the feeding days. Furthermore, 1 child was excluded, because it was not fully clear when she had received a blood transfusion. Age, anthropometric features, Hb, PF and CRP concentrations of the children at baseline are summarized in **Table 1**. Stunting (height-for-age Z-score < –2) was observed in 18 children (~35%); 3 (~6%) of them were severely stunted. Eight of the children (~15%) were underweight (weight-for-age Z-score < – 2 > –3) and 3 (~6%) were wasted (weight for-height Z-score < –2 > –3). Twenty-two children (~42%) were anemic. ID and ID anemia were found in 9 (~17%) and 5 (~10%) children, respectively. Ten children (~19%) had CRP concentrations >5 mg/L at baseline.

Eight of the children (~15%) had a positive stool smear for *Schistosoma mansoni* and 3 children (~6%) had detected hookworms at baseline and/or endpoint measurements. According to the infection intensities defined by WHO (25), 1 child had a moderate-intensity-infection of *S. mansoni* while all the other children had low-intensity-infections of *S. mansoni* or hookworms. The infected children were treated against soil transmitted helminthes after the endpoint measurements. Twelve (~23%)

and 6 (~12%) children had asymptomatic malaria at baseline and at endpoint measurements, respectively. These children were treated against malaria parasitemia after the endpoint measurements.

Table 1: Age, anthropometric features, and Hb, PF, and CRP concentrations of participating young Beninese children[1]

Variable	Study 1	Study 2	Study 3
Participants, n	18	16	18
Age, mo	28.2 ± 5.3	30.8 ± 4.8	26.9 ± 3.9
Weight, kg	10.8 ± 1.1	11.7 ± 1.3	10.9 ± 1.2
Height, cm	84.2 ± 4.8	85.3 ± 4.6	86.2 ± 4.8
Height-for-age Z-score	−1.85 ± 0.65[b]	−1.95 ± 0.89[b]	−0.91 ± 1.35[a]
Weight-for-height Z-score	−0.56 ± 0.66[a,b]	−0.09 ± 0.71[a]	−1.06 ± 0.79[b]
Weight-for-age Z-score	−1.38 ± 0.66	−1.16 ± 0.62	−1.23 ± 0.89
Hb, g/L	109 ± 10	110 ± 8	109 ± 8
PF, µg/L	21.8 (14.8, 32.3)	21.9 (15.6, 30.7)	28.6 (21.1, 38.8)
Plasma CRP, mg/L	1.28 (0.60, 2.74)	1.28 (0.58, 2.81)	1.04 (0.51, 2.10)

[1] Values are means ± SDs or geometric means (95% CIs) analyzed as log-transformed data. Labeled means in a row without a common letter differ, $P < 0.05$. CRP, C-reactive protein; Hb, hemoglobin; PF, plasma ferritin.

Test meal composition

The native iron concentration in the millet porridge was 4.5 ± 0.2 mg/100 g dry matter (DM) and in the CFF was 1.4 ± 0.0 mg/100 g fresh matter (FM). The phytate and polyphenol concentrations in the millet porridge were 376 ± 1 mg/100 g DM and 8 ± 0 mg GAEs/100 g FM, respectively. Phytate and polyphenol concentrations in the CFF were 568 ± 16 mg/100 g FM and 80 ± 1 mg GAEs/100 g FM, respectively. The meal based on 80 g millet porridge (5.5% DM) blended with 30 g CFF therefore contained 0.7 mg native iron, 6 mg fortificant iron, 191 mg phytate, and 32 mg polyphenols. The phytate:iron molar ratio in the iron-fortified test meals was 2.5:1. The molar ratio of AA to iron in the test meals with extra AA was 3.8:1 compared with 1.9:1 with standard but no extra AA.

Table 2: Fractional iron absorption and total absorbed iron per test meal consumed by young Beninese children[1]

Study	n	Meal	Iron fortification (label)	Meal additives	Fractional iron absorption	Total absorbed iron per meal
					% of dose	mg
1	18	1A	3 mg ^{58}FeSO$_4$ + 3 mg ^{56}FeSO$_4$	—	8.4 (5.6, 12.6)	0.56 (0.38, 0.84)
		1B	3 mg ^{57}FeSO$_4$ + 3 mg Na^{57}FeEDTA	—	5.9 (3.8, 9.0)*	0.39 (0.25, 0.60)*
2	16	2A	3 mg ^{58}FeSO$_4$ + 3 mg ^{56}FeSO$_4$	—	7.3 (4.8, 11.1)	0.49 (0.32, 0.74)
		2B	3 mg ^{54}FeSO$_4$ + 3 mg ^{56}FeSO$_4$	AA	11.6 (7.8, 17.2)**	0.77 (0.52, 1.14)**
3	18	3A	3 mg ^{58}FeSO$_4$ + 3 mg ^{56}FeSO$_4$	—	8.0 (5.5, 11.5)^c	0.53 (0.37, 0.77)^c
		3B	3 mg ^{57}FeSO$_4$ + 3 mg ^{56}FeSO$_4$	Phytase	15.8 (11.1, 22.6)^b	1.05 (0.74, 1.50)^b
		3C	3 mg ^{54}FeSO$_4$ + 3 mg ^{56}FeSO$_4$	Phytase, AA	19.9 (14.4, 27.6)^a	1.33 (0.96, 1.83)^a

[1] Values are geometric means (95% CIs) analyzed as log-transformed data. Asterisks indicate different from meal A: *$P < 0.01$; **$P < 0.001$. Labeled means within a study without a common letter differ, $P < 0.05$. All meals consisted of 80 g millet porridge blended with 30 g CFF and 40 mg AA. Meals 2B and 3C contained 40 mg extra AA. AA, ascorbic acid; CFF, complementary food fortificant.

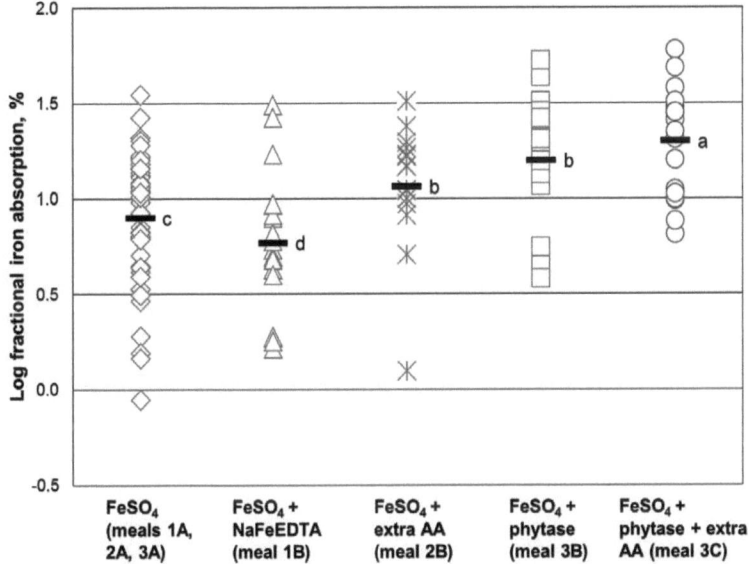

Figure 1: Between-study comparison of fractional iron absorption from different types of test meals consumed by young Beninese children. Values are log-transformed individual data points with the horizontal bar representing the log-transformed mean. Fractional absorption data from the same type of meal administered in the different studies (meals 1A, 2A and 3A) were combined for statistical analysis. Labeled means without a common letter differ, $P < 0.05$. Test meals were based on 80 g millet porridge blended with 30 g lipid-based CFF and 40 mg ascorbic acid. AA, ascorbic acid; CFF, complementary food fortificant.

Iron absorption measurements

As shown in **Table 2**, fractional mean iron absorption in study 1 from the meals fortified with $FeSO_4$ was ~30% higher than that from the test meals fortified with a mixture of NaFeEDTA and $FeSO_4$ ($P < 0.01$). In study 2, the addition of extra AA (40 mg) increased the fractional iron absorption from $FeSO_4$ ($P < 0.001$). In study 3, adding phytase almost doubled fractional iron absorption from $FeSO_4$-fortified test meals ($P < 0.001$). The addition of extra AA to the test meals containing phytase increased fractional iron absorption compared with the test meals containing phytase without extra AA ($P < 0.05$). Adding AA and/or phytase to the meals increased the geometric mean of total absorbed iron per meal to >0.6 mg, which is the needed amount of daily iron in children 12–36 mo of age. When adding phytase and extra AA, the total absorbed iron more than doubled. A between-study comparison (Fig. 1) showed no difference in fractional absorption when phytase was added compared with adding extra AA ($P > 0.05$). Iron absorption when adding phytase plus extra AA was higher than adding only extra AA without phytase ($P < 0.05$). Fractional absorption from $FeSO_4$-fortified test meals containing phytase with or without extra AA was higher than that from test meals fortified with a mixture of $FeSO_4$ and NaFeEDTA ($P < 0.05$).

Discussion

In study 2, the addition of extra AA increased the fractional iron absorption by ~60%. The enhancing effect of AA on human nonheme iron absorption has been reported in several previous studies (26, 27), and is related to the reducing and chelating properties of AA during digestion of the food (28). The 40 mg AA chosen as the standard concentration to meet AA requirements in this study is between the 30 mg proposed by FAO/WHO for healthy children aged 1–5 y old (29) and the 70–90 mg proposed in a recent publication for moderate, acute, malnourished children of the same age group (30). We chose 40 mg AA primarily because the CFF should be a preventive rather than a therapeutic complementary food and also because 40 mg resulted in an AA:iron molar ratio of ~2:1, which is recommended to enhance iron absorption from foods with normal phytate concentrations (31). Our results, however, indicate that a higher AA:iron molar ratio of 4:1 in the CFF would further enhance iron absorption. AA is easily oxidized to inactive compounds by exposure to air and to some degree by heat treatment (28); therefore, further studies are needed to ensure adequate AA levels in CFF after storage.

It was rather surprising that fractional iron absorption from test meals fortified with the mixture of NaFeEDTA and $FeSO_4$ was significantly lower than from test meals fortified with only $FeSO_4$. A potential explanation for this effect is that the standard concentration of AA added in our study increased the absorption of $FeSO_4$ but not that from NaFeEDTA. Previous studies reported no benefit to nonheme iron absorption when AA was added to NaFeEDTA-fortified test meals at AA:iron molar ratios of 5:1 (32) and 3:1 (33); thus, it is possible that the relatively high amount of AA added in our study enhanced absorption from $FeSO_4$ but not from NaFeEDTA, resulting in a lower overall absorption from NaFeEDTA.

EDTA is an iron chelator and binds iron strongly at the pH of gastric juice. After pH rises in the duodenum, EDTA exchanges ferric iron for other metals, thus releasing iron for absorption. EDTA is thus reported to prevent iron from forming insoluble, nonabsorbable complexes in the stomach with food components such as phytate and to release iron from the iron-EDTA complex in the duodenum for absorption (31). NaFeEDTA or the EDTA moiety alone (added as Na_2EDTA) have repeatedly been reported to enhance iron absorption from phytate-containing meals in adults (34, 35) and also in children (36, 37). However, in less inhibitory meals, the use of NaFeEDTA or Na_2EDTA did not enhance iron absorption (38, 39). In sugar cane syrup, which contained no phytate or other known iron absorption inhibitors, iron absorption from NaFeEDTA was 67% lower than from $FeSO_4$ (40) and earlier work indicated that increasing the molar ratio of EDTA to iron with Na_2EDTA progressively decreased iron absorption (41). Cook and Monsen (41) investigated the effect of EDTA on iron absorption from a composite American meal and concluded that the inhibitory effect of EDTA on iron absorption can only be excluded at EDTA to iron molar ratios <0.5 (41). Additionally, results in infants have not always been conclusive (42).

In the present study, the total amount of native and fortificant iron in the test meals was 6.7 mg (3 mg iron as NaFeEDTA), resulting in a molar ratio of EDTA to iron of ~0.5:1. This molar ratio is in the range that has been reported to increase iron absorption in adults from phytate-containing cereal meals (35, 39). Our test meal was only partly based on cereals, and contained high amounts of fat, low to moderate amounts of phytate and a high amount of AA. In such a meal, the potential benefit of NaFeEDTA may be differentially affected by the EDTA to iron molar ratios. Similar to our study, Chang et al. (43) investigated a NaFeEDTA-fortified complementary food partly based on cereals. Compared with $FeSO_4$-fortified meals, they reported no change in absorption in infants at a molar ratio of EDTA to iron of 0.7:1 but a significant increase in iron absorption at a molar ratio of 0.4:1.

Another factor that could have decreased iron absorption from NaFeEDTA in the present study is *Helicobacter pylori*. The moderately malnourished children who participated in the study live in an area where *H. pylori* infections are widely prevalent (44, 45). *H. pylori* infections are associated with an increased postprandial duodenal acid load and a prolongation of low pH in the duodenal bulb (46, 47). This could possibly impair iron release from EDTA, because there is no rise in pH that weakens the bond between EDTA and iron.

Based on our findings, we cannot recommend NaFeEDTA as an iron fortificant for an LNS or CFF. It is more expensive than ferrous sulfate (48) and, in the presence of recommended AA concentrations (29, 31), offers no improvement in absorption. Another disadvantage is the limited quantity of

NaFeEDTA that can be added because of the ADI (14). Further studies are needed to investigate the reasons for the lower iron absorption from NaFeEDTA in our study.

The addition of microbial phytase just before consumption in study 3 almost doubled iron absorption compared with the control meal. This increase is similar to 2 previous human studies using a phytase active at gastric pH in highly inhibitory meals based on whole maize (32) or on wheat bran (49). The increase in our study was slightly higher than in the previous study using the same phytase in whole-maize meals (32). This was probably due to the lower amount of phytate in our test meal and due to the addition of more FTUs. Adding extra AA together with phytase had an additive effect and further increased iron absorption compared with standard AA concentrations. This is comparable to earlier findings where iron absorption from dephytinized, AA-fortified, wheat-soy porridges was higher than from the dephytinized porridges without AA (50).

LNSs have been widely used in the field of child malnutrition and were well accepted by children and caregivers (12, 51, 52). Current LNSs are typically made from vegetable oil, peanut paste, milk powder, sugar, and a micronutrient premix. The use of milk powder and canola oil contributes to the reasonably high costs of LNSs at the present time. High costs might limit the long-term use of LNSs among the most vulnerable population groups. Replacing milk powder and canola oil with domestic products such as defatted soybean flour or soybean oil would make LNSs cheaper and they could be locally manufactured (12). On the down side, soybean flour will introduce more phytate into the LNSs. In our study, we used canola oil, but it could be easily replaced by soybean oil, which is not expected to be different from canola oil in terms of affecting iron absorption. Soybean oil is readily available in many African countries and the essential fatty acid composition is only slightly inferior to canola oil (12).

To our knowledge, the present study is the first study in young children investigating iron absorption from an LNS type using phytase. Our study provides important insights for the optimization of iron absorption from LNSs designed to prevent ID in young children. Our findings suggest that adding phytase to a CFF based on locally available products and blended with cereal porridge would be a very promising food-based approach to simultaneously improve bioavailable iron and energy intake of young children. However, several questions have to be solved. In our study, the phytase was added just before consumption. For a commercial CFF, the phytase would be optimally mixed and packaged with the other ingredients and stability to storage must be confirmed. In a similar way to AA, phytase is heat sensitive; thus, mothers would need to be instructed when to add the phytase-containing CFF to the porridge. As an important step towards regulatory approval, the Joint WHO/FAO expert committee on Food Additives allocated an ADI "not specified" for use of *A. niger* phytase in applications such as LNSs, FBFs and MNPs (53). An alternative to phytase would be extra AA or, even better, a mixture of phytase and extra AA. Our study was a single-meal study and may not accurately predict long-term absorption. However, we think that a CFF optimized for iron absorption is a promising approach to increase bioavailable iron and energy intake in children living in a malaria-endemic area.

Acknowledgements

The authors thank Zuckermühle Rupperswill AG and Florin AG for providing ingredients free of charge for the CFF production. A special thank you goes to Valid International (Oxford, UK), which helped elaborating the formula of CFF. The authors also express thanks to DSM Nutritional Products, Switzerland for donating the phytase. C.I.C., I.M.E., C.Z., E.M., J.D.H., and R.F.H. designed research; C.I.C., J.H., and F.T. conducted research; C.I.C. and C.Z. analyzed data; C.I.C., I.M.E., and R.F.H. wrote the paper; and C.I.C., I.M.E., and R.F.H. had primary responsibility for final content. All authors read and approved the final manuscript.

References

1. WHO/UNICEF/UNU. Iron deficiency anemia: assessment, prevention and control. Geneva: World Health Organization, 2001.
2. Dewey KG. Increasing iron intake of children through complementary foods. Food Nutr Bull. 2007;28:S595-609.
3. Gibson RS, Ferguson EL, Lehrfeld J. Complementary foods for infant feeding in developing countries: their nutrient adequacy and improvement. Eur J Clin Nutr. 1998;52:764-70.
4. WHO. Complementary Feeding of Young Children in Developing Countries: A Review of Current Scientific Knowledge. Geneve: World Health Organization, 1998.
5. de Pee S, Bloem MW. Current and potential role of specially formulated foods and food supplements for preventing malnutrition among 6-to 23-month-old children and for treating moderate malnutrition among 6-to 59-month-old children. Food Nutr Bull. 2009;30:434-63.
6. Dewey K, Berger J, Chen JS, Chen CM, de Pee S, Huffman S, Kraemer K, Lartey A, Lutter C, Maleta K, et al. Formulations for fortified complementary foods and supplements: Review of successful products for improving the nutritional status of infants and young children. Food Nutr Bull. 2009;30:239-55.
7. Adu-Afarwuah S, Lartey A, Brown KH, Zlotkin S, Briend A, Dewey KG. Randomized comparison of 3 types of micronutrient supplements for home fortification of complementary foods in Ghana: effects on growth and motor development. Am J Clin Nutr. 2007;86:412-20.
8. Adu-Afarwuah S, Lartey A, Brown KH, Zlotkin S, Briend A, Dewey KG. Home fortification of complementary foods with micronutrient supplements is well accepted and has positive effects on infant iron status in Ghana. Am J Clin Nutr. 2008;87:929-38.
9. Collins S. Changing the way we address severe malnutrition during famine. Lancet. 2001;358:498-501.
10. Diop el HI, Dossou NI, Ndour MM, Briend A, Wade S. Comparison of the efficacy of a solid ready-to-use food and a liquid, milk-based diet for the rehabilitation of severely malnourished children: a randomized trial. Am J Clin Nutr. 2003;78:302-7.
11. Manary MJ, Ndekha MJ, Ashorn P, Maleta K, Briend A. Home based therapy for severe malnutrition with ready-to-use food. Arch Dis Child. 2004;89:557-61.
12. Phuka J, Ashorn U, Ashorn P, Zeilani M, Cheung YB, Dewey KG, Manary M, Maleta K. Acceptability of three novel lipid-based nutrient supplements among Malawian infants and their caregivers. Matern Child Nutr. 2011;7:368-77.
13. Walczyk T, Davidsson L, Zavaleta N, Hurrell RF. Stable isotope labels as a tool to determine the iron absorption by Peruvian school children from a breakfast meal. Fresen J Anal Chem. 1997;359:445-9.
14. European Food Safety Authority. Scientific opinion on the use of ferric sodium EDTA as a source of iron added for nutritional purposes to foods for the general population (including food supplements) and to food for particular nutritional uses. Parma: European Food Safety Authority; 2010.

15. Troesch B, van Stujivenberg ME, Smuts CM, Kruger HS, Biebinger R, Hurrell RF, Baumgartner J, Zimmermann MB. A micronutrient powder with low doses of highly absorbable iron and zinc reduces iron and zinc deficiency and improves weight-for-age z-scores in South african children. J Nutr. 2011;141:237-42.
16. Makower RU. Extraction and Determination of Phytic Acid in Beans (Phaseolus-Vulgaris). Cereal Chem. 1970;47:288-96.
17. Van Veldhoven PP, Mannaerts GP. Inorganic and Organic Phosphate Measurements in the Nanomolar Range. Anal Biochem. 1987;161:45-8.
18. Singleton VL, Orthofer R, Lamuela-Raventos RM. Analysis of total phenols and other oxidation substrates and antioxidants by means of Folin-Ciocalteu reagent. In: Packer L, editor. Oxidants and antioxidants, Pt A. San Diego: Academic Press; 1999. p.152–78.
19. Daland GA, Castle WB. A Simple and Rapid Method for Demonstrating Sickling of the Red Blood Cells - the Use of Reducing Agents. J Lab Clin Med. 1948;33:1082-8.
20. Hotz K, Krayenbuehl PA, Walczyk T. Mobilization of storage iron is reflected in the iron isotopic composition of blood in humans. J Biol Inorg Chem. 2012;17:301-9.
21. Cercamondi CI, Egli IM, Ahouandjinou E, Dossa R, Zeder C, Salami L, Tjalsma H, Wiegerinck E, Tanno T, Hurrell RF, et al. Afebrile Plasmodium falciparum parasitemia decreases absorption of fortification iron but does not affect systemic iron utilization a double stable-isotope study in young Beninese women. Am J Clin Nutr. 2010;92:1385-92.
22. Linderkamp O, Versmold HT, Riegel KP, Betke K. Estimation and prediction of blood volume in infants and children. Eur J Pediatr. 1977;125:227-34.
23. Turnlund JR, Keyes WR, Peiffer GL. Isotope Ratios of Molybdenum Determined by Thermal Ionization Mass-Spectrometry for Stable-Isotope Studies of Molybdenum Metabolism in Humans. Anal Chem. 1993;65:1717-22.
24. Rios E, Hunter RE, Cook JD, Smith NJ, Finch CA. The absorption of iron as supplements in infant cereal and infant formulas. Pediatrics. 1975;55:686-93.
25. WHO. Prevention and control of schistosomiasis and soil-transmitted helminthiasis. Geneva: World Health Organization, 2002.
26. Fidler MC, Davidsson L, Zeder C, Walczyk T, Marti I, Hurrell RF. Effect of ascorbic acid and particle size on iron absorption from ferric pyrophosphate in adult women. Int J Vitam Nutr Res. 2004;74:294-300.
27. Hallberg L, Brune M, Rossander L. Effect of ascorbic acid on iron absorption from different types of meals. Studies with ascorbic-acid-rich foods and synthetic ascorbic acid given in different amounts with different meals. Hum Nutr Appl Nutr. 1986;40:97-113.
28. Hurrell RF, Lynch S, Bothwell T, Cori H, Glahn R, Hertrampf E, Kratky Z, Miller D, Rodenstein M, Streekstra H, et al. Enhancing the absorption of fortification iron. A SUSTAIN Task Force report. Int J Vitam Nutr Res. 2004;74:387-401.
29. FAO/WHO/UNU. Human energy requirements. Report of a joint FAO/WHO expert consultation. Rome: FAO, 2001.
30. Golden MH. Proposed recommended nutrient densities for moderately malnourished children. Food Nutr Bull. 2009;30:267-342.
31. Hurrell RF. Fortification: Overcoming technical and practical barriers. J Nutr. 2002;132:806-12.
32. Troesch B, Egli I, Zeder C, Hurrell RF, de Pee S, Zimmermann MB. Optimization of a phytase-containing micronutrient powder with low amounts of highly bioavailable iron for in-home fortification of complementary foods. Am J Clin Nutr. 2009;89:539-44.
33. Macphail AP, Bothwell TH, Torrance JD, Derman DP, Bezwoda WR, Charlton RW, Mayet F. Factors Affecting the Absorption of Iron from Fe(III)EDTA. Brit J Nutr. 1981;45:215-27.
34. International Nutritional Anemia Consultative Group (INACG). Iron EDTA for food fortification. New York: Nutrition Foundation, 1993.
35. Hurrell RF, Reddy MB, Burri J, Cook JD. An evaluation of EDTA compounds for iron fortification of cereal-based foods. Brit J Nutr. 2000;84:903-10.
36. Davidsson L, Dimitriou T, Boy E, Walczyk T, Hurrell RF. Iron bioavailability from iron-fortified Guatemalan meals based on corn tortillas and black bean paste. Am J Clin Nutr. 2002;75:535-9.
37. Davidsson L, Walczyk T, Zavaleta N, Hurrell R. Improving iron absorption from a Peruvian school breakfast meal by adding ascorbic acid or Na2EDTA. Am J Clin Nutr. 2001;73:283-7.
38. Fidler MC, Davidsson L, Walczyk T, Hurrell RF. Iron absorption from fish sauce and soy sauce fortified with sodium iron EDTA. Am J Clin Nutr. 2003;78:274-8.

39. Macphail AP, Patel RC, Bothwell TH, Lamparelli RD. EDTA and the Absorption of Iron from Food. Am J Clin Nutr. 1994;59:644-8.
40. Martinez-Torres C, Romano EL, Renzi M, Layrisse M. Fe(III)-EDTA complex as iron fortification. Further studies. Am J Clin Nutr. 1979;32:809-16.
41. Cook JD, Monsen ER. Food iron absorption in man II. The effect of EDTA on absorption of dietary non-heme iron. Am J Clin Nutr. 1976;29:614-20.
42. Davidsson L, Ziegler E, Zeder C, Walczyk T, Hurrell R. Sodium iron EDTA [NaFe(III)EDTA] as a food fortificant: erythrocyte incorporation of iron and apparent absorption of zinc, copper, calcium, and magnesium from a complementary food based on wheat and soy in healthy infants. Am J Clin Nutr. 2005;81:104-9.
43. Chang SY, Huang ZW, Ma YX, Piao JH, Yang XG, Zeder C, Hurrell RF, Egli I. Mixture of ferric sodium ethylenediaminetetraacetate (NaFeEDTA) and ferrous sulfate: An effective iron fortificant for complementary foods for young Chinese children. Food Nutr Bull. 2012;33:111-6.
44. Holcombe C, Omotara BA, Eldridge J, Jones DM. H. pylori, the most common bacterial infection in Africa: a random serological study. Am J Gastroenterol. 1992;87:28-30.
45. Kidd M, Louw JA, Marks IN. Helicobacter pylori in Africa: observations on an 'enigma within an enigma'. J Gastroenterol Hepatol. 1999;14:851-8.
46. Hamlet A, Olbe L. The influence of Helicobacter pylori infection on postprandial duodenal acid load and duodenal bulb pH in humans. Gastroenterology. 1996;111:391-400.
47. Olbe L, Fandriks L, Hamlet A, Svennerholm AM, Thoreson AC. Mechanisms involved in Helicobacter pylori induced duodenal ulcer disease:an overview. World J Gastroenterol. 2000;6:619-23.
48. Bothwell TH, MacPhail AP. The potential role of NaFeEDTA as an iron fortificant. Int J Vitam Nutr Res. 2004;74:421-34.
49. Sandberg AS, Hulthen LR, Turk M. Dietary Aspergillus niger phytase increases iron absorption in humans. J Nutr. 1996;126:476-80.
50. Hurrell RF, Reddy MB, Juillerat MA, Cook JD. Degradation of phytic acid in cereal porridges improves iron absorption by human subjects. Am J Clin Nutr. 2003;77:1213-9.
51. Hess SY, Bado L, Aaron GJ, Ouedraogo JB, Zeilani M, Brown KH. Acceptability of zinc-fortified, lipid-based nutrient supplements (LNS) prepared for young children in Burkina Faso. Matern Child Nutr. 2011;7:357-67.
52. Adu-Afarwuah S, Lartey A, Zeilani M, Dewey KG. Acceptability of lipid-based nutrient supplements (LNS) among Ghanaian infants and pregnant or lactating women. Matern Child Nutr. 2011;7:344-56.
53. WHO. Evaluation of certain food additives: seventy-sixth report of the Joint FAO/WHO Expert Committee on Food Additives. Geneva: World Health Organization, 2012.

Online Supporting Material

$$R_1 = \frac{{}^{56}h_{nat} \cdot n_{nat} + {}^{56}h_A \cdot n_A + {}^{56}h_B \cdot n_B + {}^{56}h_C \cdot n_C}{{}^{54}h_{nat} \cdot n_{nat} + {}^{54}h_A \cdot n_A + {}^{54}h_B \cdot n_B + {}^{54}h_C \cdot n_C} \quad (1)$$

$$R_2 = \frac{{}^{56}h_{nat} \cdot n_{nat} + {}^{56}h_A \cdot n_A + {}^{56}h_B \cdot n_B + {}^{56}h_C \cdot n_C}{{}^{57}h_{nat} \cdot n_{nat} + {}^{57}h_A \cdot n_A + {}^{57}h_B \cdot n_B + {}^{57}h_C \cdot n_C} \quad (2)$$

$$R_3 = \frac{{}^{56}h_{nat} \cdot n_{nat} + {}^{56}h_A \cdot n_A + {}^{56}h_B \cdot n_B + {}^{56}h_C \cdot n_C}{{}^{58}h_{nat} \cdot n_{nat} + {}^{58}h_A \cdot n_A + {}^{58}h_B \cdot n_B + {}^{58}h_C \cdot n_C} \quad (3)$$

The total circulating iron is the sum of the circulating natural iron and of the circulating isotopic labels:

$$n_{nat} + n_A + n_B + n_C = N_{tot} \quad (4)$$

Equations (1), (2), (3) and (4) can be rearranged in the following form:

$(R_1 \cdot {}^{54}h_{nat} - {}^{56}h_{nat}) \cdot n_{nat} + (R_1 \cdot {}^{54}h_A - {}^{56}h_A) \cdot n_A + (R_1 \cdot {}^{54}h_B - {}^{56}h_B) \cdot n_B + (R_1 \cdot {}^{54}h_C - {}^{56}h_C) \cdot n_C = 0$

$(R_2 \cdot {}^{57}h_{nat} - {}^{56}h_{nat}) \cdot n_{nat} + (R_2 \cdot {}^{57}h_A - {}^{56}h_A) \cdot n_A + (R_2 \cdot {}^{57}h_B - {}^{56}h_B) \cdot n_B + (R_2 \cdot {}^{57}h_C - {}^{56}h_C) \cdot n_C = 0$

$(R_3 \cdot {}^{58}h_{nat} - {}^{56}h_{nat}) \cdot n_{nat} + (R_3 \cdot {}^{58}h_A - {}^{56}h_A) \cdot n_A + (R_3 \cdot {}^{58}h_B - {}^{56}h_B) \cdot n_B + (R_3 \cdot {}^{58}h_C - {}^{56}h_C) \cdot n_C = 0$

$n_{nat} + n_A + n_B + n_C = N_{tot}$

After substitution (i.e. a_{11} for $(R_1 \cdot {}^{54}h_{nat} - {}^{56}h_{nat})$), the set of equations can be written as:

$a_{11} n_{nat} + a_{12} n_A + a_{13} n_B + a_{14} n_C = b_1$
$a_{21} n_{nat} + a_{22} n_A + a_{23} n_B + a_{24} n_C = b_2$
$a_{31} n_{nat} + a_{32} n_A + a_{33} n_B + a_{34} n_C = b_3$
$a_{41} n_{nat} + a_{42} n_A + a_{43} n_B + a_{44} n_C = b_4$

Supplemental Figure 1: Calculation of iron absorption. The measured isotopic ratios R_i can be expressed in the form of equations (1), (2) and (3), where n_m are the amounts in mol of the natural iron (m=nat), the ^{54}Fe (m=A), ^{57}Fe (m=B) and ^{58}Fe (m=C) labels circulating in the blood, respectively, and $^j h_m$ is the isotopic abundance in % mol in the natural iron and the isotopic labels, with j being the mass number of the respective iron isotope.

Manuscript 4 – Iron-Biofortified Millet Study

Total iron absorption by young women from iron-biofortified pearl millet composite meals is double that from regular millet meals, but less than that from post-harvest iron-fortified millet meals

Colin I Cercamondi[1], Ines M Egli[1], Evariste Mitchikpe[2], Felicien Tossou[3], Christophe Zeder[1], Joseph D Hounhouigan[2] and Richard F Hurrell[1]

[1] Laboratory of Human Nutrition, Institute of Food, Nutrition and Health, ETH Zurich, Zurich, Switzerland; [2] Laboratoire de Physiologie de la Nutrition, Université d'Abomey Calavi, Benin; [3] Zone Sanitaire Natitingou, Ministère de la Santé, Benin

Journal of Nutrition 2013; 143: 1376–82

Supported by HarvestPlus and the INSTAPA project, which receives funding from the European Community's Seventh Framework Programme [FP7/2007–2013] under grant agreement no. 211484

Abstract

Iron biofortification of pearl millet (*Pennisetum glaucum*) is a promising approach to combat iron deficiency (ID) in the millet-consuming communities of developing countries. To evaluate the potential of iron-biofortified millet to provide additional bioavailable iron compared with regular millet and post-harvest iron-fortified millet, an iron absorption study was conducted in 20 Beninese women with marginal iron status. Composite test meals consisting of millet paste based on regular-iron, iron-biofortified, or post-harvest iron-fortified pearl millet flour accompanied by a leafy vegetable sauce or an okra sauce were fed as multiple meals for 5 d. Iron absorption was measured as erythrocyte incorporation of stable iron isotopes. Fractional iron absorption from test meals based on regular-iron millet (7.5%) did not differ from iron-biofortified millet meals (7.5%; $P = 1.0$), resulting in a higher quantity of total iron absorbed from the meals based on iron-biofortified millet (1125 µg vs. 527 µg; $P < 0.0001$). Fractional iron absorption from post-harvest iron-fortified millet meals (10.4%) was higher than from regular-iron and iron-biofortified millet meals ($P < 0.05$ and $P < 0.01$, respectively), resulting in a higher quantity of total iron absorbed from the post-harvest iron-fortified millet meals (1500 µg; $P < 0.0001$ and $P < 0.05$, respectively). Results indicate that consumption of iron-biofortified millet would double the amount of iron absorbed and, although fractional absorption of iron from biofortification is less than that from fortification, iron-biofortified millet should be highly effective in combatting ID in millet-consuming populations.

Introduction

Biofortification, which refers to the development of micronutrient-enhanced staple crop varieties by traditional breeding practices or by modern biotechnology, has gained increased attention in preventing micronutrient deficiencies over the last decade (1). It is potentially more sustainable and cost-effective than conventional fortification and it implicitly targets the low-income households in remote areas with large daily consumption of a few food staples and limited access to commercially marketed fortified foods (2–4). Dissemination of seeds that efficiently accumulate soil iron could increase the delivery of iron to the diets of poverty-stricken people, including women and children who are most at risk for iron deficiency (ID) (4). However, iron biofortification only improves iron status if the additional iron provided by the biofortified crop is bioavailable and consequently fills the gap between current iron intake and iron requirement. Furthermore, acceptance of biofortified crop varieties by farmers and consumers is crucial. This implies that the biofortified crop has a sufficiently high yield that is stable in different environments and climatic zones and that the cooking and sensory properties are comparable to nonbiofortified varieties (4, 5).

Present iron biofortification research programs are focused on enhancing iron concentrations in pearl millet, maize, wheat, rice, and beans (1, 6, 7). Pearl millet (*Pennisetum glaucum*) is among the most important staple crops in the semi-arid tropics of India (8) and sub-Saharan Africa, especially Western Africa (9), where 26% of the average per capita cereal grain consumption has been reported to be pearl millet (10). The monotonous pearl millet-based diets with low iron concentration and low iron bioavailability contribute to ID with or without anemia in these regions (11).

Depending on genotype and environmental conditions, the iron concentration of pearl millet has been reported to vary between 1.6 and 9.6 mg/100 g (12, 13). Several studies have reported higher concentrations up to 20 mg iron/100 g pearl millet (14–16); however, such concentrations most likely include contaminant iron from post-harvest treatments and should not be used for reporting native iron concentration of pearl millet (17). The iron concentration of iron-biofortified pearl millet has been reported to be ~7–8 mg/100 g (18, 19), which is about double the iron content of other major cereal staples. Compared with regular pearl millet, the iron-biofortified varieties usually also have higher phytic acid (PA) concentrations (18). PA is a well-known inhibitor of human iron absorption (20, 21), impairing the bioavailability of additional iron in biofortified varieties (22) as well as iron fortification compounds used for conventional fortification (23).

Several previous studies have reported improved iron status through conventional iron fortification of cereals flours (24). However, pearl millet has not been considered as a vehicle for iron fortification so far, most likely due to the limited industrial processing of pearl millet grains in sub-Saharan Africa and India (10). Therefore, iron-biofortified millet providing additional bioavailable iron could be a promising approach to combat ID in pearl millet-consuming communities with limited access to commercially fortified foods. The aim of the present study was to investigate the extent to which the additional iron in iron-biofortified pearl millet is bioavailable. Fractional iron absorption and total iron absorbed from composite meals based on iron-biofortified pearl millet were compared with the same composite meals consisting of regular-iron pearl millet or post-harvest iron-fortified pearl millet using a 5-d multiple meal design and stable isotope technique.

Methods

Participants

The study was carried out in young Beninese women recruited in Natitingou, Atacora department, Northern Benin. Twenty-two apparently healthy nonpregnant, nonlactating women with marginal iron stores [plasma ferritin (PF) <25 µg/L], aged between 17 and 35 y and having a body weight <65 kg were selected from an initial screening of 133 women. Intake of vitamin and/or mineral supplements was not allowed during and 2 wk before the study. Women with symptomatic malaria (asexual *P. falciparum* parasitemia in blood smears + fever); known metabolic, chronic, and gastro-intestinal disease; as well as women on long-term medication were excluded from the study. No women were recruited who had donated blood or experienced substantial blood loss within 6 mo of the beginning of the study. All study participants provided informed written consent. Ethical approval for the study was given by the ethical review committee at the Ministry of Health in Benin and at the ETH Zurich, Switzerland.

Study design

A randomized cross-over design with multiple meals was used with each woman serving as her own control. Every woman received 3 different types of test meals in series of 10 servings for 5 d each. The order of the 3 different series was randomized. The 3 different types of test meals were either based on regular-iron, iron-biofortified, or post-harvest iron-fortified pearl millet (**Fig. 1**). Servings of one test meal type were always labeled with the same isotope: ^{54}Fe was used for the post-harvest iron-fortified meals, ^{57}Fe for the regular-iron meals, and ^{58}Fe for the iron-biofortified meals. The labeled test meals were administered twice per day in the morning and at noon from Wednesday to Sunday for 3 consecutive weeks (d 1–5, d 8–12, and d 15–19). The serving in the morning was administered between 0630 and 0930 h after an overnight fast and the second serving was administered at least 3 h later. The participants consumed the test meals completely in the presence of the investigators and were not allowed to eat or drink between the test meals and for 3 h after the second meal.

During screening (baseline measurements), 2 wk before the first meal feeding series, body weight and height of the participants were measured and a first blood sample was drawn for iron status determination [hemoglobin (Hb), PF, C-reactive protein (CRP)] and malaria parasitemia diagnosis. Stool and urine samples were taken for the detection of soil transmitted helminths and a pregnancy test. A second (d 1), third (d 8), and fourth (d 15) blood sample was drawn for iron status (Hb, PF, CRP) and malaria parasitemia determination immediately before starting a meal feeding series. Fourteen days after the last test meal (d 33; endpoint measurements), body weight and height measures were repeated and a fifth blood sample was drawn for iron isotopic and malaria parasitemia analysis. Iron absorption was determined by using stable isotope technique in which the incorporation into erythrocytes of isotopic iron labels was measured 14 d after the administration of the last test meal (13). Soil transmitted helminth analyses and pregnancy tests from the screening were confirmed by stool and urine samples taken at endpoint. Ear temperature was measured at screening and endpoint and always before the meal feedings.

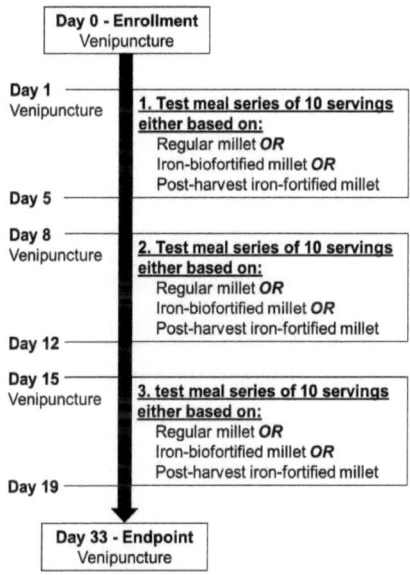

Figure 1: Schematic diagram of the study design.

Test meals

Test meals were composite meals of traditional Beninese millet paste served either with a leafy vegetables sauce in the morning or an okra sauce at noon. Bottled water (300 g) was administered in 2 servings of 80 and 220 g with each test meal. Each test meal serving consisted of 60 g pearl millet flour prepared into millet paste (325 ± 2 g/serving) accompanied by either 110 ± 2 g of leafy vegetable sauce or 80 ± 2 g okra sauce. The 2 sauces were prepared freshly for each study day according to a standardized procedure adapted from local recipes. Recipes of the 2 sauces are shown in **Supplemental Table 1**.

Nongenetically modified iron-biofortified (ICTP8203) and regular-iron (DG-9444) pearl millet (*P. glaucum*) was planted and harvested by

HarvestPlus India and then shipped to Benin as whole grains. A portable household mill (HAWOS Billy 200; Hawos Kornmühlen) was used to obtain the flour for the millet paste preparation. Millet paste was prepared freshly on each study day by weighing and mixing millet flour and water at a ratio of 1.2:1. This blend was then added to boiling water (blend:boiling water ~1:2.3) and cooked for 26 ± 3 min with intermittent stirring.

Regular-iron, iron-biofortified and post-harvest iron-fortified millet meals were labeled with 0.4 mg ^{57}Fe, 0.4 mg ^{58}Fe, and 0.4 mg ^{54}Fe, respectively. The stable iron isotopes were in form of a solution and were diluted in the first serving of water (80 g) and administered after one-half of the millet paste and sauce was consumed. To ensure complete intake of isotopic labels, the second serving of water (220 g) was served in the same plastic tumbler. Ferrous sulfate solution (4 g/L) was used for the post-harvest fortification of regular-iron millet. The iron concentration per serving of regular-iron millet meals was adjusted to approximately the same concentration as in the iron-biofortified millet meals. To prevent potential organoleptic changes in the meals, the necessary amount of iron, 0.9 ml ferrous sulfate solution (4 g Fe/L), was diluted in the same 80 g of water as the isotopic labels and administered after one-half of the millet paste and sauce was consumed.

Test meal analysis

Iron concentrations in the millet flours, millet pastes, and sauces were analyzed by graphite-furnace atomic absorption spectrophotometry (GF-AAS, AA240Z) after mineralization by microwave digestion (MLS ETHOSplus; MLS) using a mixture of HNO_3 and H_2O_2. The PA concentration was measured by using a modification of the Makower method (25) in which iron was replaced by cerium in the precipitation step.

After the mineralization of the precipitates, inorganic phosphate was determined according to Van Veldhoven and Mannaerts (26) and converted into PA concentrations. The total polyphenol (PP) concentration was determined by using a modification of the Folin-Ciocalteau method (27) and was expressed as gallic acid equivalents. As sauces were prepared daily with freshly bought vegetables, iron, PA, and PP concentrations of each preparation (15/sauce) were measured and expressed as means ± SDs.

Preparation of isotopically labeled iron

Isotopically-labeled $^{54}FeSO_4$, $^{57}FeSO_4$, and $^{58}FeSO_4$ were prepared from isotopically enriched elemental iron (^{54}Fe-metal: 99.9 % enriched; ^{57}Fe-metal: 97.9 % enriched; ^{58}Fe-metal: 99.9 % enriched; all Chemgas) by dissolution in 0.1 mol/L sulfuric acid. The solutions were flushed with argon to keep the iron in the +II oxidation state. Prepared iron tracer solutions were analyzed for iron isotopic composition and tracer iron concentration by reversed isotope dilution MS.

Blood Analysis

Hb was measured in whole blood on the day of collection by using HemoCue hemoglobin 201+; anemia was defined as Hb <12 g/dL (28). PF and CRP were measured using an IMMULITE automatic system (Siemens). ID was defined as PF <15 µg/L and ID anemia as Hb <12 g/dL plus PF <15 µg/L (28). The expected high-sensitivity CRP concentrations for healthy individuals were <5 mg/L (29). Thick and thin blood smears were stained in duplicate by using the Giemsa coloration technique and were examined independently by 2 experienced microscopists (30).

Each isotopically enriched blood sample was analyzed in duplicate for its isotopic composition. Whole blood was mineralized by microwave digestion and iron was separated by anion-exchange chromatography and a subsequent precipitation step with ammonium hydroxide (31). Iron isotope ratios were determined by a multicollector inductively coupled plasma MS instrument (NEPTUNE; Thermo Finnigan).

Stool and urine analysis

Each stool sample was analyzed in duplicate for the detection of soil-transmitted helminths using the KatoKatz method (30). Urine samples were collected for a pregnancy test (HcG distinct rapid test device; Ziva Impex) and the detection of *Schistosoma haematobium* eggs using the syringe filtration technique (30).

Calculation of Fe absorption

The amounts of ^{54}Fe, ^{57}Fe, and ^{58}Fe labels in the blood were calculated based on the shift in iron isotope ratios and the estimated amount of iron circulating in the body. Circulating iron was calculated based on the blood volume estimated from height and weight and measured Hb concentration (32). The calculations were based on the principles of isotope dilution, taking into account that iron isotopic labels were not monoisotopic, using the methods described by Turnlund et al. (33). Calculation of iron absorption is shown in detail in **Supplemental Figure 1**. For calculation of fractional absorption, 80% incorporation of the absorbed Iron into RBCs was assumed (34).

Statistical analysis

Analyses were conducted with SPSS statistical software (SPSS 16.0) and Excel (Windows 7; Microsoft). Results of food analysis, age, anthropometric features, Hb, PF, and CRP were presented as means ± SDs if normally distributed. If not normally distributed, the results were presented as geometric mean values with the 95% CI in parentheses. Results of iron absorption were presented as geometric mean values with the 95% CI in parentheses. Iron absorption from different test meals within the same participant was compared by repeated-measures ANOVA followed by a Bonferroni corrected pairwise comparison. Comparison of millet flour composition (iron, PP, PA) was done by Mann-Whitney U tests. Differences were considered significant at $P < 0.05$. All data were converted to their logarithms for statistical analysis and reconverted for reporting. The study was powered to detect an intra-subject difference of 30% in fractional iron absorption with an α level of 0.05.

Results

Participant characteristics

The data of 20 women were included in the study. The data of one woman were excluded due to impaired health conditions during the whole study, and data of another woman were excluded due to high CRP concentration (49.3 mg/L) measured before the second meal-feeding series. At baseline, all women had marginal iron status (PF <25 µg/L) and normal CRP concentrations (<5 mg/L) (**Table 1**). Five of the women were iron deficient without anemia, 10 were iron-deficient anemic, and 1 woman was anemic at baseline.

During the study, PF concentrations between 25 and 32 µg/L were measured in 2 women before 1 of the 3 meal-feeding series. Additionally, 2 women had PF concentrations between 25 and 32 µg/L before 2 of the 3 meal-feeding series. One additional woman had PF concentrations between 25 and 30 µg/L prior to all the 3 meal-feeding series. Three women had slightly elevated CRP concentrations (5.5–7.5 mg/L) before

1 of the 3 meal-feeding series. None of the women had a CRP concentration >5 mg/L before more than one meal-feeding series. All women were negative for soil transmitted helminths at baseline, but at endpoint, 3 women were positive for *S. mansoni*. None of the women had a positive malaria blood smear at baseline or endpoint.

Table 1: Age, anthropometric features, and Hb, PF, and CRP concentrations of Beninese women at baseline[1]

Variable	Summary value
Age, *y*	20.6 ± 2.9
Weight, *kg*	54.2 ± 6.2
Height, *cm*	161 ± 7
BMI, *kg/m²*	20.9 ± 2.6
Hb, *g/L*	119 ± 13
PF, *µg/L*	11.9 ± 5.1
Plasma CRP, *mg/L*	0.51 (0.33, 0.88)

[1] Values are means ± SDs or geometric means (95% CIs), n = 20. CRP, C-reactive protein; Hb, hemoglobin; PF, plasma ferritin.

Test meal composition

The iron concentration of the iron-biofortified millet was ~3.5 times that of regular-iron millet (**Table 2**). Depending on the sauce, the final iron concentrations of the composite test meals were ~50-60% higher in meals based on iron-biofortified and post-harvest iron-fortified millet than in meals based on regular-iron millet (**Table 3**). The iron concentrations of the leafy vegetable sauce and the okra sauce were 2.0 ± 0.5 and 1.2 ± 0.2 mg/100 g fresh matter (FM), respectively. The PA concentration in the iron-biofortified millet flour was ~200 mg/100 g higher than that in the regular-iron millet flour (Table 2), resulting in a difference of ~120 mg PA/serving of millet paste (Table 3). The PA concentrations of the leafy vegetable sauce (2 ± 1 mg/100 g FM) and okra sauce (10 ± 4 mg/100 g

FM) were low and had no relevant influence on the PA:iron (PA:iron molar ratio) in the test meals. PA:iron was highest in the regular-iron millet meals followed by the iron-biofortified millet meals and lowest in the post-harvest iron-fortified millet meals (Table 3). The PP concentrations in the 2 millet types were similar (Table 2). The leafy vegetable sauce and okra sauce had PP concentrations of 145 ± 24 and 93 ± 9 mg gallic acid equivalents/100 g FM, respectively.

Table 2: Total PPs, PA, and iron in regular-iron and iron-biofortified pearl millet flour[1]

Pearl millet	PP	PA	Iron	PA:iron
	mg/100 g flour			
Regular-iron pearl millet (DG-9444)	106 ± 4^a	653 ± 17	2.5 ± 0.1^b	22.1:1
Iron-biofortified pearl millet (ICPT8203)	87 ± 1^b	852 ± 35	8.8 ± 0.3^a	8.2:1

[1] Values are means $\pm$ SDs or molar ratios, $n = 3$. Labeled means in a column without a common letter differ, $P < 0.05$. PA, phytic acid; PA:iron, phytic acid: iron molar ratio; PP, polyphenol.

Iron absorption measurements

Mean fractional iron absorption from the iron-biofortified millet meals did not differ ($P = 1.0$) compared with the regular-iron millet meals (**Table 4**), resulting in a doubling of total iron absorbed from the iron-biofortified meals ($P < 0.0001$). The mean fractional absorption from the post-harvest iron-fortified millet meals was ~40% higher than from the iron-biofortified millet meals ($P < 0.01$) and the regular-iron millet meals ($P < 0.05$). Total iron absorbed from the test meals based on post-harvest iron-fortified millet was therefore higher than from the regular-iron and iron-biofortified millet meals ($P < 0.0001$ and $P < 0.05$, respectively).

Table 3: Total PP, PA, and iron in millet pastes and composite millet meals consumed by Beninese women[1]

Pearl millet meal	PP	PA	Iron[2]	PA:iron[3]
		mg/serving		
Regular-iron millet paste	72 ± 2	392 ± 10	1.5 ± 0.2	
Composite millet meal[4]				
+ leafy vegetable sauce	231 ± 27	394 ± 10	4.1 ± 0.5	8.1:1
+ okra sauce	146 ± 7	400 ± 10	2.9 ± 0.3	11.7:1
Iron-biofortified millet paste	65 ± 1	511 ± 21	5.5 ± 0.6	
Composite millet meal[4]				
+ leafy vegetable sauce	224 ± 27	513 ± 21	8.1 ± 0.8	5.4:1
+ okra sauce	139 ± 7	519 ± 21	6.9 ± 0.6	6.4:1
Post-harvest iron-fortified millet paste	72 ± 2	392 ± 10	1.5 ± 0.2[5]	
Composite millet meal[4]				
+ leafy vegetable sauce	231 ± 27	394 ± 10	7.8 ± 0.5	4.3:1
+ okra sauce	146 ± 7	400 ± 10	6.6 ± 0.3	5.1:1

[1] Values are means ± SDs, $n = 3$. PA, phytic acid; PA:iron, phytic acid:iron molar ratio; PP, polyphenol.
[2] Iron values of the millet pastes include only native iron. Iron values of the composite millet meals include native iron and 0.4 mg iron as ^{54}Fe, ^{57}Fe or ^{58}Fe. Post-harvest iron-fortified composite millet meals contained 3.7 mg iron added as $^{56}FeSO_4$.
[3] Values are molar ratios of the composite millet meals.
[4] Values are means ± SDs based on the means from the analysis of single components (pastes, $n = 3$; sauces, $n = 15$). SDs were adapted by calculating the square root of the sum from the square of the SDs from the single analysis of pastes and sauces. Iron, PA and PP concentrations of the sauces alone are in text.
[5] Value does not include 3.7 mg fortification iron which was added later on to the composite millet meal.

Table 4: Fractional and total iron absorption per composite millet meals consumed by Beninese women[1]

Composite millet meal	Fractional iron absorption	Total iron absorption[2]	Ratio of fractional absorption (meal A:meal B, C)
	% of dose	*mg/d*	
Regular-iron millet meal	7.5 (5.7, 10.0)[b]	0.53 (0.40, 0.70)[c]	—
Iron-biofortified millet meal	7.5 (5.6, 10.1)[b]	1.13 (0.83, 1.52)[b]	1.0
Post-harvest iron-fortified millet meal	10.4 (8.2, 13.2)[a]	1.50 (1.18, 1.91)[a]	0.7

[1] Values are geometric means (95% CIs), $n = 20$. Labeled means in a column without a common letter differ, $P < 0.05$.
[2] Total iron absorption is based on the fractional absorption and the iron concentrations of one portion of millet paste with leafy vegetable sauce and one portion of millet paste with okra sauce. Iron concentrations of the test meals are shown in Table 3.

Discussion

The current study has 2 major findings. The first is that fractional absorption did not differ between the regular-iron and the iron-biofortified millet meals, leading to a significantly increased quantity of total iron absorbed from the iron-biofortified millet meals compared with regular-iron millet meals. This indicates that iron-biofortified pearl millet would provide higher amounts of bioavailable iron than regular-iron pearl millet when consumed in a composite meal. Our findings with iron-biofortified millet differ from a previous study comparing iron bioavailability from regular and iron-biofortified beans. The iron-biofortified beans did not provide a greater amount of absorbed iron when administered in multiple composite meals for 5 d (22). The authors argued that the higher PA concentrations in the iron-biofortified beans compared with the regular-iron beans led to a molar excess of PA and therefore more strongly inhibited the bioavailability of additional iron irrespective of the PA:iron, which was around 9:1 in both types of bean meals. The simultaneous increase of iron and PA in iron-biofortified crops has been reported in previous studies (18, 22) and makes it difficult for plant breeders to develop iron-biofortified crops providing bioavailable iron. Our results suggest that, in contrast to common beans, the additional iron in iron-biofortified pearl millet is not strongly inhibited by the additional PA in iron-biofortified millet. This is most probably because the PA concentrations in the 2 millet types used in our study were generally lower than in the bean varieties, and because the difference in total PA between the regular-iron and iron-biofortified millet was only ~200 mg/100 g compared with ~500 mg/100 g between the regular-iron and iron-biofortified beans. Furthermore, the difference in iron concentrations between the regular-iron and iron-biofortified millet was greater than in the beans.

Although the PA:iron in the iron-biofortified millet meals was lower than in the regular-iron millet meals, fractional iron absorption was the same from both meal types. This can be explained by a relative decrease in fractional iron absorption with increased quantity of ingested iron. Cook et al. (35) added 1, 3, and 5 mg labeled $FeSO_4$ to a bread roll meal and reported that fractional iron absorption decreases as the amount of iron ingested increases but that more iron was absorbed from the meals with higher iron concentrations.

The second finding of the present study is that the post-harvest iron-fortified millet meals provided a greater amount of absorbed iron than the iron-biofortified millet meals. The fortification iron added to the regular-iron millet meals raised the iron concentration to that of the iron-biofortified millet meals. The lower PA concentration in regular-iron millet compared with iron-biofortified millet led to the lower PA:iron in the post-harvest iron-fortified millet meals. The lower PA:iron (below <6:1) could explain why the fractional iron absorption from post-harvest iron-fortified millet meals is higher than from the other 2 meal types. It has been reported that iron absorption from composite meals improves with PA:iron <6:1 (36).

The composite meals used in our study were very close to the traditional Beninese preparations of pearl millet paste accompanied by a leafy vegetable sauce or okra sauce. However, if available and affordable, local people like to add meat, fish, or traditional Beninese cheese to the sauces. We did not use these ingredients, because it would have introduced heme iron or proteins which might have interfered with iron absorption. We did not measure ascorbic acid concentrations in the 2 sauces but assume negligible amounts of ascorbic acid after cooking (37). PP concentrations in the test meals were relatively high. Some PPs are potent inhibitors of iron absorption (38); however, the relatively high fractional iron absorptions in our study indicate little interference of PPs with iron absorption. Unlike sorghum, pearl millet does not contain condensed tannins with catechol and galloyl groups (39), which are suspected to inhibit iron absorption (40). We also assume that the sauces did not contain many PPs with iron-binding structures. *Afitin*, an indispensable fermented Beninese condiment (41), can contain considerable amounts of PA (42); however, we used reduced amounts of a long-fermented *afitin* (24 h) to prepare our sauces. Moreover, the standardized sauces in our study did not contain ingredients that would have added a considerable amount of PA, such as peanut or pumpkin seed paste. The use of ingredients or sauces high in PA would affect the amount of bioavailable iron from iron-biofortified millet as it would that of regular-iron millet. However, we cannot completely rule out if ingredients high in PA would not add PA in molar excess and bind the additional iron from the iron-biofortified millet and thus reduce its benefit compared with regular-iron millet.

The daily per capita consumption of millet in women 18–45 y of age from rural Northern Benin is 159 g (43). Hundred fifty-nine g of iron-biofortified millet prepared into millet paste with 7.5% iron bioavailability, as measured in our study, would provide ~72% of the median daily iron requirements for menstruating women older than 18 y (44). The same amount of regular-iron millet would only provide ~20% of the requirements (calculations do not include iron from sauces). Millet consumption data for young Beninese children are not available but data from Burkina Faso, which borders to Northern Benin, can be extrapolated. Assuming equal fractional iron absorption in young children, Beninese children 12–36 mo of age, who consume an average of 32 g millet/d (45), could satisfy ~46% of the 0.5 mg absorbed iron required per day (44) with iron-biofortified millet consumption but only ~13% if the regular-iron millet is consumed.

An early radioisotope study measuring iron bioavailability from composite pearl millet meals reported iron absorption of 4.8 and 1.2%, respectively, from a pearl millet couscous meal with fish and a pearl millet gruel meal with peanut paste. The study calculated the iron absorption by iron-replete men with an average serum ferritin of 83 µg/L in relation to exchangeable iron, because the meals were contaminated by iron from soil residues and/or by dust (46). In our study, fractional iron absorption values from meals based on regular-iron millet were higher than those reported in the previous study and relatively high for meals with such high PA:iron. The reason for this finding is probably the upregulated iron absorption in our mainly iron-deficient female participants (47).

In conclusion, our study shows that the total amount of iron absorbed from iron-biofortified and post-harvest iron-fortified pearl millet is about 2- and 3-times higher than from regular-iron pearl millet. The PA:iron appeared to be the major determinant of the total amount of iron absorbed. Our results suggest that, despite delivering higher PA concentrations, biofortification of pearl millet could be a valuable approach to increase the bioavailable iron supply in remote millet-consuming communities with limited access to conventionally post-harvest fortified foods. Efficacy studies are now needed to investigate if iron-biofortified millet provides sufficient additional iron to improve iron status and combat ID in such populations. Although our study does not exactly represent conventional iron fortification, we think that the results can be extrapolated to the absorption of fortification iron, because we assume that the iron in the aqueous solution consumed with the meal enters the common exchangeable iron pool where its bioavailability is influenced by enhancers and inhibitors in the same way as the bioavailability of an iron premix added to the millet flour before preparation. $FeSO_4$ would be a suitable iron compound in terms of bioavailability, but further research is needed in relation to sensory changes. Although fractional iron absorption is somewhat better from iron-fortified than iron-biofortified millet, post-harvest fortification of pearl millet could be more challenging because of the lack of central milling facilities or because of the difficulties in fortifying flour at the community level. We therefore conclude that iron biofortification of pearl millet is a promising approach to easily and rapidly increase bioavailable iron in the diets of millet-consuming communities in Western Africa and probably also in India, depending on the type of foods it is consumed with.

Acknowledgements

The authors thank HarvestPlus, especially Erick Boy, for providing the regular-iron and iron-biofortified pearl millet.

C.I.C., I.M.E., C.Z., E.M., J.D.H., and R.F.H. designed research; C.I.C., E.M., and F.T. conducted research; C.I.C. and C.Z. analyzed data; C.I.C., I.M.E., and R.F.H. wrote the paper; and C.I.C., I.M.E., and R.F.H. had primary responsibility for final content. All authors read and approved the final version of the paper.

References

1. Nestel P, Bouis HE, Meenakshi JV, Pfeiffer W. Biofortification of staple food crops. J Nutr. 2006;136:1064-7.
2. Meenakshi JV, Johnson NL, Manyong VM, Degroote H, Javelosa J, Yanggen DR, Naher F, Gonzalez C, Garcia J, Meng E. How Cost-Effective is Biofortification in Combating Micronutrient Malnutrition? An Ex ante Assessment. World Dev. 2010;38:64-75.
3. Qaim M, Stein AJ, Meenakshi JV. Economics of biofortification. Agr Econ-Blackwell. 2007;37:119-33.

4. Bouis HE, Hotz C, McClafferty B, Meenakshi JV, Pfeiffer WH. Biofortification: A new tool to reduce micronutrient malnutrition. Food Nutr Bull. 2011;32:S31-S40.
5. Bouis HE, Welch RM. Biofortification-A Sustainable Agricultural Strategy for Reducing Micronutrient Malnutrition in the Global South. Crop Sci. 2010;50:S20-S32.
6. Gregorio GB. Progress in breeding for trace minerals in staple crops. J Nutr. 2002;132:500s-2s.
7. Ortiz-Monasterio JI, Palacios-Rojas N, Meng E, Pixley K, Trethowan R, Pena RJ. Enhancing the mineral and vitamin content of wheat and maize through plant breeding. J Cereal Sci. 2007;46:293-307.
8. Parthasarathy Rao P, Birthal PS, Reddy BVS, Rai KN, Ramesh S. Diagnostics of Sorghum and Pearl Millet Grains-based Nutrition in India. Journal of SAT agricultural research 2006 [cited 2013 July]; Available from: http://ejournal.icrisat.org/cropimprovement/v2i1/v2i1diagnostics.pdf
9. Vietmeyer ND, Ruskin FR. Lost Crops of Africa: Grains. Washington DC: National Academy Press, 1996.
10. Ndjeunga J, Nelson CH. Prospects for a Pearl Millet and Sorghum Food Processing Industry in West Africa Semi-Arid Tropics. In: Towards Sustainable Sorghum Production, Utilization, and Commercialization in West and Central Africa Proceedings of a Technical Workshop of the West and Central Africa Sorghum Research Network. 1999 [cited January 2013]; Available from: http://oar.icrisat.org/4891/
11. Zimmermann MB, Chaouki N, Hurrell RF. Iron deficiency due to consumption of a habitual diet low in bioavailable iron: a longitudinal cohort study in Moroccan children. Am J Clin Nutr. 2005;81:115-21.
12. Barikmo I, Ouattara F, Oshaug A. Differences in micronutrients content found in cereals from various parts of Mali. J Food Compos Anal. 2007;20:681-7.
13. Lestienne I, Buisson M, Lullien-Pellerin V, Picq C, Treche S. Losses of nutrients and anti-nutritional factors during abrasive decortication of two pearl millet cultivars (Pennisetum glaucum). Food Chem. 2007;100:1316-23.
14. Abdalla AA, El Tinay AH, Mohamed BE, Abdalla AH. Effect of traditional processes on phytate and mineral content of pearl millet. Food Chem. 1998;63:79-84.
15. Badau MH, Nkama I, Jideani IA. Phytic acid content and hydrochloric acid extractability of minerals in pearl millet as affected by germination time and cultivar. Food Chem. 2005;92:425-35.
16. Ragaee S, Abdel-Aal EM, Noaman M. Antioxidant activity and nutrient composition of selected cereals for food use. Food Chem. 2006;98:32-8.
17. Kayode APP, Linnemann AR, Nout MJR, Van Boekel MAJS. Impact of sorghum processing on phytate, phenolic compounds and in vitro solubility of iron and zinc in thick porridges. J Sci Food Agric. 2007;87:832-8.
18. Hama F, Icard-Verniere C, Guyot JP, Rochette I, Diawara B, Mouquet-Rivier C. Potential of non-GMO biofortified pearl millet (Pennisetum glaucum) for increasing iron and zinc content and their estimated bioavailability during abrasive decortication. Int J Food Sci Technol. 2012;47:1660-8.
19. HarvestPlus. Iron pearl millet. 2009 [cited October 2012]; Available from: http://www.unscn.org/layout/modules/resources/files/HarvestPlus_Pearl_Millet_Strategy.pdf
20. Hallberg L, Brune M, Rossander L. Iron-Absorption in Man - Ascorbic-Acid and Dose-Dependent Inhibition by Phytate. Am J Clin Nutr. 1989;49:140-4.
21. Hurrell RF, Juillerat MA, Reddy MB, Lynch SR, Dassenko SA, Cook JD. Soy protein, phytate, and iron absorption in humans. Am J Clin Nutr. 1992;56:573-8.
22. Petry N, Egli I, Gahutu JB, Tugirimana PL, Boy E, Hurrell R. Stable Iron Isotope Studies in Rwandese Women Indicate That the Common Bean Has Limited Potential as a Vehicle for Iron Biofortification. J Nutr. 2012;142:492-7.
23. Hurrell RF. Phytic acid degradation as a means of improving iron absorption. Int J Vitam Nutr Res. 2004;74:445-52.
24. Hurrell R, Ranum P, de Pee S, Biebinger R, Hulthen L, Johnson Q, Lynch S. Revised recommendations for iron fortification of wheat flour and an evaluation of the expected impact of current national wheat flour fortification programs. Food Nutr Bull. 2010;31:S7-21.
25. Makower RU. Extraction and Determination of Phytic Acid in Beans (Phaseolus-Vulgaris). Cereal Chem. 1970;47:288-96.
26. Van Veldhoven PP, Mannaerts GP. Inorganic and Organic Phosphate Measurements in the Nanomolar Range. Anal Biochem. 1987;161:45-8.

27. Singleton VL, Orthofer R, Lamuela-Raventos RM. Analysis of total phenols and other oxidation substrates and antioxidants by means of Folin-Ciocalteu reagent. In: Parker L, editor. Oxidants and Antioxidants, Pt A. San Diego: Academic Press; 1999. p. 152-78.
28. World Health Organization. Iron deficiency anemia: assessment, prevention and control. Geneva: WHO, 2001.
29. Dati F, Schumann G, Thomas L, Aguzzi F, Baudner S, Bienvenu J, Blaabjerg O, BlirupJensen S, Carlstrom A, HyltoftPetersen P, et al. Consensus of a group of professional societies and diagnostic companies on guidelines for interim reference ranges for 14 proteins in serum based on the standardization against the IFCC/BCR/CAP reference material (CRM 470). Eur J Clin Chem Clin. 1996;34:517-20.
30. Cercamondi CI, Egli IM, Ahouandjinou E, Dossa R, Zeder C, Salami L, Tjalsma H, Wiegerinck E, Tanno T, Hurrell RF, et al. Afebrile Plasmodium falciparum parasitemia decreases absorption of fortification iron but does not affect systemic iron utilization a double stable-isotope study in young Beninese women. Am J Clin Nutr. 2010;92:1385-92.
31. Hotz K, Krayenbuehl PA, Walczyk T. Mobilization of storage iron is reflected in the iron isotopic composition of blood in humans. J Biol Inorg Chem. 2012;17:301-9.
32. Brown E, Hopper J, Jr., Hodges JL, Jr., Bradley B, Wennesland R, Yamauchi H. Red cell, plasma, and blood volume in the healthy women measured by radiochromium cell-labeling and hematocrit. J Clin Invest. 1962;41:2182-90.
33. Turnlund JR, Keyes WR, Peiffer GL. Isotope Ratios of Molybdenum Determined by Thermal Ionization Mass-Spectrometry for Stable-Isotope Studies of Molybdenum Metabolism in Humans. Anal Chem. 1993;65:1717-22.
34. Hosain F, Marsaglia G, Noyes W, Finch CA. The nature of internal iron exchange in man. Transactions of the Association of American Physicians. 1962;75:59-63.
35. Cook JD, Minnich V, Moore CV, Rasmusse.A, Bradley WB, Finch CA. Absorption of Fortification Iron in Bread. Am J Clin Nutr. 1973;26:861-72.
36. Hurrell R, Egli I. Iron bioavailability and dietary reference values. Am J Clin Nutr. 2010;91:1461S-7S.
37. Somsub W, Kongkachuichai R, Sungpuag P, Charoensiri R. Effects of three conventional cooking methods on vitamin C, tannin, myo-inositol phosphates contents in selected Thai vegetables. J Food Compos Anal. 2008;21:187-97.
38. Hurrell RF, Reddy M, Cook JD. Inhibition of non-haem iron absorption in man by polyphenolic-containing beverages. Brit J Nutr. 1999;81:289-95.
39. Dykes L, Rooney LW. Sorghum and millet phenols and antioxidants. J Cereal Sci. 2006;44:236-51.
40. Brune M, Hallberg L, Skanberg AB. Determination of Iron-Binding Phenolic Groups in Foods. J Food Sci. 1991;56:128-&.
41. Azokpota P, Hounhouigan DJ, Nago MC. Microbiological and chemical changes during the fermentation of African locust bean (Parkia biglobosa) to produce afitin, iru and sonru, three traditional condiments produced in Benin. Int J Food Microbiol. 2006;107:304-9.
42. Avallone S, Bohuon P, Hemery Y, Treche S. Improvement of the in vitro digestible iron and zinc content of okra (Hibiscus esculentus L.) sauce widely consumed in Sahelian Africa. J Food Sci. 2007;72:S153-S8.
43. Ategbo EAD. Food and Nutrition Insecurity in Northern Benin: Impact on Growth Performance of Children and on Year to Year Nutritional Status of Adults. Wageningen: Wageningen University; 1993.
44. FAO/WHO/UNU. Vitamin and mineral requirements in human nutrition. Report of a joint FAO/WHO expert consultation. 2nd edition ed. Bangkok: FAO/WHO, 2004.
45. Hama F. Rétention et biodisponibilité du fer et du zinc au cours des procédés de préparation des plats traditionnels à base de céréales locales ou biofortifiées, consommés par les jeunes enfants au Burkina Faso. Montpellier: Université Montpellier 2; 2012.
46. Guiro AT, Galan P, Cherouvrier F, Sall MG, Hercberg S. Iron-Absorption from African Pearl-Millet and Rice Meals. Nutr Res. 1991;11:885-93.
47. Zimmermann MB, Biebinger R, Egli I, Zeder C, Hurrell RF. Iron deficiency up-regulates iron absorption from ferrous sulphate but not ferric pyrophosphate and consequently food fortification with ferrous sulphate has relatively greater efficacy in iron-deficient individuals. Brit J Nutr. 2011;105:1245-50.

Online Supporting Material

Supplemental Table 1: Recipes of the two sauces accompanying the millet paste[1]

Ingredients	Leafy vegetable sauce[2]	Okra sauce[3]
	g/kg	
African eggplant-leaves (*Solanum macrocarpon*)	396	102
Okra (*Abelmoschus esculentus*)	-	153
Mallow-leaves (*Corchorus olitorius*)	-	88
Tomato	170	-
Onion	85	26
Red palm oil	71	31
Afitin[4]	57	29
Salt	6	7
Green chili	6	4
Maggi Cube	6	4
Black pepper	1	1
Bicarbonate	1	3
Water	202	551

[1] Sauces were prepared freshly the day before administration and stored overnight in a refrigerator.
[2] 110 g of leafy vegetable sauce accompanied the serving of pearl millet paste in the morning.
[3] 80 g of okra sauce accompanied the serving of pearl millet paste at noon.
[4] Fermented paste of African locust bean (*Parkia biglobosa*) (41).

Online Supporting Material

$$R_1 = \frac{{}^{56}h_{nat} \cdot n_{nat} + {}^{56}h_A \cdot n_A + {}^{56}h_B \cdot n_B + {}^{56}h_C \cdot n_C}{{}^{54}h_{nat} \cdot n_{nat} + {}^{54}h_A \cdot n_A + {}^{54}h_B \cdot n_B + {}^{54}h_C \cdot n_C} \quad (1)$$

$$R_2 = \frac{{}^{56}h_{nat} \cdot n_{nat} + {}^{56}h_A \cdot n_A + {}^{56}h_B \cdot n_B + {}^{56}h_C \cdot n_C}{{}^{57}h_{nat} \cdot n_{nat} + {}^{57}h_A \cdot n_A + {}^{57}h_B \cdot n_B + {}^{57}h_C \cdot n_C} \quad (2)$$

$$R_3 = \frac{{}^{56}h_{nat} \cdot n_{nat} + {}^{56}h_A \cdot n_A + {}^{56}h_B \cdot n_B + {}^{56}h_C \cdot n_C}{{}^{58}h_{nat} \cdot n_{nat} + {}^{58}h_A \cdot n_A + {}^{58}h_B \cdot n_B + {}^{58}h_C \cdot n_C} \quad (3)$$

The total circulating iron is the sum of the circulating natural iron and of the circulating isotopic labels:

$$n_{nat} + n_A + n_B + n_C = N_{tot} \quad (4)$$

Equations (1), (2), (3) and (4) can be rearranged in the following form:

$(R_1 \cdot {}^{54}h_{nat} - {}^{56}h_{nat}) \cdot n_{nat} + (R_1 \cdot {}^{54}h_A - {}^{56}h_A) \cdot n_A + (R_1 \cdot {}^{54}h_B - {}^{56}h_B) \cdot n_B + (R_1 \cdot {}^{54}h_C - {}^{56}h_C) \cdot n_C = 0$

$(R_2 \cdot {}^{57}h_{nat} - {}^{56}h_{nat}) \cdot n_{nat} + (R_2 \cdot {}^{57}h_A - {}^{56}h_A) \cdot n_A + (R_2 \cdot {}^{57}h_B - {}^{56}h_B) \cdot n_B + (R_2 \cdot {}^{57}h_C - {}^{56}h_C) \cdot n_C = 0$

$(R_3 \cdot {}^{58}h_{nat} - {}^{56}h_{nat}) \cdot n_{nat} + (R_3 \cdot {}^{58}h_A - {}^{56}h_A) \cdot n_A + (R_3 \cdot {}^{58}h_B - {}^{56}h_B) \cdot n_B + (R_3 \cdot {}^{58}h_C - {}^{56}h_C) \cdot n_C = 0$

$n_{nat} + n_A + n_B + n_C = N_{tot}$

After substitution (i.e. a_{11} for $(R_1 \cdot {}^{54}h_{nat} - {}^{56}h_{nat})$), the set of equations can be written as:

$a_{11} n_{nat} + a_{12} n_A + a_{13} n_B + a_{14} n_C = b_1$
$a_{21} n_{nat} + a_{22} n_A + a_{23} n_B + a_{24} n_C = b_2$
$a_{31} n_{nat} + a_{32} n_A + a_{33} n_B + a_{34} n_C = b_3$
$a_{41} n_{nat} + a_{42} n_A + a_{43} n_B + a_{44} n_C = b_4$

Supplemental Figure 1: Calculation of iron absorption. The measured isotopic ratios R_i can be expressed in the form of equations (1), (2) and (3), where n_m are the amounts in mol of the natural iron (m=nat), the ^{54}Fe (m=A), ^{57}Fe (m=B) and ^{58}Fe (m=C) labels circulating in the blood, respectively, and $^j h_m$ is the isotopic abundance in % mol in the natural iron and the isotopic labels, with j being the mass number of the respective iron isotope.

CONCLUSIONS AND PERSPECTIVES

The overall aim of the thesis was to develop approaches to improve iron nutrition of sorghum and millet consuming communities in endemic malarial areas. Sorghum and millets are main staple foods for the majority of the most poverty-stricken people in sub-Saharan Africa and in some drier areas of India (Vietmeyer and Ruskin 1996). Despite their importance and their substantial production levels, sorghum and millet have been neglected by the nutrition community. They are still crops of the small cultivator and the development of a sorghum- or millet-based food-processing industry has remained in its early stages (Sanders and Ouendeba 2012). By investigating iron bioavailability from fortified sorghum and millet foods in malaria endemic areas, this thesis generated important information for future iron (bio)fortification programs with sorghum and millets. The new knowledge will help to develop fortified sorghum- and millet-based foods which provide sufficient bioavailable iron to combat the high prevalence of iron deficiency (ID) in young children and women of childbearing age in sub-Saharan Africa and can also be applied to other semi-arid tropic areas such as India.

Endemic malaria parasitemia affects human iron homeostasis by several mechanisms (Lamikanra et al. 2007) and is prevalent in regions where sorghum and millets are the major staple crops. Previous studies have indicated that parasitemia and even asymptomatic infections, which are very common in endemic areas, may negatively affect human iron absorption (Doherty et al. 2008; de Mast et al. 2010). In this thesis, the study in Beninese women showed a ~40% reduction in fractional iron absorption during asymptomatic malaria affirming for the first time the inhibition of iron absorption by asymptomatic malarial parasitemia. The study shows that the decrease in iron absorption is mediated through low-grade inflammation during asymptomatic malaria. Such a decrease may contribute to ID and possibly blunt the efficacy of fortification programs. Further research is needed to investigate whether asymptomatic malaria has a long-term effect on iron absorption from fortified foods, which then would require means to improve iron bioavailability or the adjustment of iron concentrations in fortified sorghum and millet foods in malaria endemic areas.

In contrast to the majority of cereals, millets and particularly sorghum can contain considerable amounts of polyphenols (PPs) (Dicko et al. 2002; Dykes and Rooney 2006). The series of absorption studies in women consuming dephytinized sorghum test meals with different PP concentrations demonstrated that brown sorghum PPs are strong inhibitors of iron absorption. The PPs from brown sorghum are mostly condensed tannins and appear to be more inhibitory than bean PPs but similarly inhibitory than PPs from herb teas. The results suggest that in contrast to other cereal foods (Hurrell et al. 2003), the significant reduction of phytic acid in foods based on brown sorghum varieties would not greatly increase iron absorption. Furthermore, the study shows that it is challenging to overcome the inhibition of PPs. NaFeEDTA only partly overcomes the inhibitory effect and a substantial reduction of PPs using polyphenol oxidase (laccase) showed no benefit on iron absorption. However, polyphenol oxidases differ in their substrate specificity and further investigations with other enzymes are required. Ascorbic acid was the most potent enhancer in the current study completely overcoming the inhibitory effect of brown sorghum PPs. However, application of ascorbic acid in iron-fortified sorghum foods would be challenging due to temperature and oxidation sensitivity of ascorbic acid. Therefore, the use of NaFeEDTA seems to be the most feasible and promising approach increasing iron bioavailability from iron-fortified sorghum foods containing both phytic acid and inhibitory PPs.

Young children are particularly vulnerable to ID due to their high iron requirements during rapid growth and development. Complementary foods in developing countries often provide insufficient amount of bioavailable iron and are often bulky cereal-based porridges with low energy density (Gibson et al. 1998). Therefore, energy-dense lipid-based nutrient supplements (LNSs) fortified with iron and other micronutrients added to traditional complementary foods, such as porridge, can simultaneously improve micronutrient and energy intake of young children. The work in this thesis demonstrated that iron bioavailability from an iron- and ascorbic acid (AA)-fortified LNS termed complementary food fortificant (CFF) blended with millet porridge can be optimized by phytase and increased concentration of AA. Surprisingly, it was also shown that iron bioavailability from a mixture of ferrous sulphate and NaFeEDTA was impaired compared to only ferrous sulphate. This suggests that AA used to fortify the CFF improves absorption from ferrous sulphate but not from NaFeEDTA. A previous study reported no beneficial effect of adding AA to a NaFeEDTA-fortified maize porridge (Troesch et al. 2009). Based on these findings and drawbacks, such as the high cost of NaFeEDTA and the restricted use due to an acceptable daily intake of EDTA, NaFeEDTA cannot be recommended as fortificant for lipid-based complementary food at the moment. Further research is needed to explain the discrepancy of absorption from NaFeEDTA used to fortify different food vehicles.

The CFF developed in this thesis is, similarly to previous LNSs, in form of a paste and contains a mixture of micro- and macronutrients which should be added as a fortificant to cereal-based porridges commonly consumed by children in developing countries. The CFF was designed as a preventive food to treat mild to moderate malnutrition. The CFF is based on ingredients that are locally available in developing countries e.g. soybean flour. Compared with previous LNSs, the soybean flour replaced the milk powder. It is cheaper than milk powder, but on the down side introduces more phytate in the CFF. In such a type of CFF containing considerable amounts of phytate, the results obtained in this thesis support the use of microbial phytase active at gastric pH in combination with increased AA concentrations to improve iron bioavailability. The best way to provide the phytase with the CFF would be to blend the phytase together with the other ingredients of the CFF. Such an approach requires research on the storage stability of phytase in a lipid-dense food matrix. Microbial phytase from *Aspergillus niger* has been approved by the Joint FAO/WHO Expert Committee on Food Additives and an ADI "not specified" was allocated to applications such as LNSs (WHO 2012).

Many of the women and young children in developing countries, vulnerable to ID, live in remote rural areas with limited access to commercially marketed fortified foods (Nestel et al. 2006). To improve the iron nutrition of such population groups, iron-biofortified crops providing additional bioavailable iron are a promising novel approach. However, iron biofortification can only be successful if the additional iron is bioavailable and if the concentration of iron in the crop after preparation provides enough iron to fill the gap between current iron intake and iron requirement. Moreover, iron-biofortified crops will only be accepted by farmers if they are stable over different environment, if they deliver the same yield as the regular crop, and if their resistance against pathogens is adequate (Bouis and Welch 2010). The study in this thesis compared the iron bioavailability from regular-iron, iron-biofortified and post-harvest iron-fortified pearl millet. The total quantity of iron absorbed from the meals based on iron-biofortified millet was significantly higher than from the regular-iron millet meals but lower than from the post-harvest iron-fortified millet meals. These results indicate that iron-biofortified and post-harvest iron-fortified pearl millet have the potential to provide additional bioavailable iron in the diets of pearl millet consuming communities compared with regular-iron pearl millet. However, post-harvest fortification of millet is challenging due to the lack of centralized milling and due to the difficulties when fortifying millet on a community-level. Therefore, in communities with limited access to conventionally fortified foods, iron-biofortified millet seems to be the most promising approach to provide additional bioavailable iron. Studies investigating the efficacy and effectiveness of iron-biofortified crops are needed to prove that the additional bioavailable iron is sufficient to improve iron status in the targeted population.

Putting all the findings together, the investigations in this thesis highlight that improving iron nutrition from sorghum and millet diets in malaria endemic areas is challenging because the negative influences of PA, PPs and malaria on iron absorption have to be considered simultaneously. Nevertheless, the thesis demonstrates that in remote areas with limited access to conventionally fortified foods, *in-home* fortification of millet-based complementary foods using enhancers, such as phytase and AA, or iron-biofortified millets have the potential to provide additional bioavailable iron into the diets of young children and women of reproductive age, respectively. However, when implementing such approaches, acceptance in the target population is crucial and the potential negative impact of malaria infections on long-term iron absorption needs further investigations.

References

Bouis, H. E. and R. M. Welch (2010). Biofortification-A Sustainable Agricultural Strategy for Reducing Micronutrient Malnutrition in the Global South. *Crop Sci* **50**(2): S20-S32.

de Mast, Q., D. Syafruddin, S. Keijmel, T. O. Riekerink, O. Deky, P. B. Asih, . . . A. J. van der Ven (2010). Increased serum hepcidin and alterations in blood iron parameters associated with asymptomatic P. falciparum and P. vivax malaria. *Haematologica* **95**(7): 1068-1074.

Dicko, M. H., R. Hilhorst, H. Gruppen, A. S. Traore, C. Laane, W. J. H. van Berkel and A. G. J. Voragen (2002). Comparison of content in phenolic compounds, polyphenol oxidase, and peroxidase in grains of fifty sorghum varieties from Burkina Faso. *J Agr Food Chem* **50**(13): 3780-3788.

Doherty, C. P., S. E. Cox, A. J. Fulford, S. Austin, D. C. Hilmers, S. A. Abrams and A. M. Prentice (2008). Iron Incorporation and Post-Malaria Anaemia. *Plos One* **3**(5): e2133.

Dykes, L. and L. W. Rooney (2006). Sorghum and millet phenols and antioxidants. *J Cereal Sci* **44**(3): 236-251.

Gibson, R. S., E. L. Ferguson and J. Lehrfeld (1998). Complementary foods for infant feeding in developing countries: their nutrient adequacy and improvement. *Eur J Clin Nutr* **52**(10): 764-770.

Hurrell, R. F., M. B. Reddy, M. A. Juillerat and J. D. Cook (2003). Degradation of phytic acid in cereal porridges improves iron absorption by human subjects. *Am J Clin Nutr* **77**(5): 1213-1219.

Lamikanra, A. A., D. Brown, A. Potocnik, C. Casals-Pascual, J. Langhorne and D. J. Roberts (2007). Malarial anemia: of mice and men. *Blood* **110**(1): 18-28.

Nestel, P., H. E. Bouis, J. V. Meenakshi and W. Pfeiffer (2006). Biofortification of staple food crops. *J Nutr* **136**(4): 1064-1067.

Sanders, J. H. and B. Ouendeba (2012). Intensive Production of Millet and Sorghum for Evolving Markets in the Sahel. Lincoln, NE, University of Nebraska.

Troesch, B., I. Egli, C. Zeder, R. F. Hurrell, S. de Pee and M. B. Zimmermann (2009). Optimization of a phytase-containing micronutrient powder with low amounts of highly bioavailable iron for in-home fortification of complementary foods. *Am J Clin Nutr* **89**(2): 539-544.

Vietmeyer, N. D. and F. R. Ruskin (1996) Lost Crops of Africa: Grains. Washington DC: National Academy Press.

WHO (2012). Evaluation of certain food additives: seventy-sixth report of the Joint FAO/WHO Expert Committee on Food Additives. Geneva, World Health Organization.

i want morebooks!

Buy your books fast and straightforward online - at one of world's fastest growing online book stores! Environmentally sound due to Print-on-Demand technologies.

Buy your books online at
www.get-morebooks.com

Kaufen Sie Ihre Bücher schnell und unkompliziert online – auf einer der am schnellsten wachsenden Buchhandelsplattformen weltweit! Dank Print-On-Demand umwelt- und ressourcenschonend produziert.

Bücher schneller online kaufen
www.morebooks.de

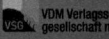

VDM Verlagsservicegesellschaft mbH
Heinrich-Böcking-Str. 6-8   Telefon: +49 681 3720 174   info@vdm-vsg.de
D - 66121 Saarbrücken   Telefax: +49 681 3720 1749   www.vdm-vsg.de

Printed by Books on Demand GmbH, Norderstedt / Germany